Beekeeping

An Ultimate Guide to Beekeeping at Home

(An Introduction to Building and Maintaining Honey Bee Colonies)

John Leister

Published By **Kate Sanders**

John Leister

All Rights Reserved

Beekeeping: An Ultimate Guide to Beekeeping at Home (An Introduction to Building and Maintaining Honey Bee Colonies)

ISBN 978-1-7776381-5-3

Legal & Disclaimer

Table Of Contents

Chapter 1: Bees And Their Importance

Bees are not just captivating creatures. They are extremely essential to our health and the environment. Below are a few of the most important reasons that bees are important to us:

1. Pollination: Bees are great pollinators and aid in the development of many varieties of plant species. Bees as pollinators make up a third of the world's crop for food. Prices and shortages of food could result if these pollinators weren't present.

2. Biodiversity Bees play a crucial contribution to the conservation of biodiversity. They help maintain the health and variety of ecosystems, by pollinating wide range of plant species. They help wildlife habitats as well as aid in the stabilization of ecosystems.

3. Value of the economy: The worldwide commercial value of pollination from bees is immense. Pollination by bees is crucial to the production of agricultural crops that amount

to millions of dollars. Bees' activities benefit the food industry, agriculture, and even the production of textiles (through the pollination of cotton).

4. Bees protect wildlife: They are not just important in the production of crops for agriculture, but in the wild plant world. A variety of species feed and rest upon these plants, creating complex ecological webs.

5. The pollination of bees is required for many medicinal plants, such as the ones used in traditional medicine. Population declines of the bee may impact the supply of the plants used to be used for health and medical purposes.

6. The production of seeds: bees aid the plants to produce seeds inside fruits. This is essential to the existence of a variety of species of plants as well as the overall health of ecosystems.

7. Ecosystem services: Bees perform essential ecosystem services that are not related to

agriculture. These include improving the health of soils as well as water retention and carbon sequestration. They help keep the natural balance and sustainability of our environment.

8. Genetic variations: colonies contain genetic variations, which helps to create more robust and productive colonies of bees. The diversity of bee populations can aid combating diseases, pests and the effects of climate changes.

9. Beauty and educational value Bees are able to teach us about the behavior of animals communicating, as well as complex social structures. With vibrant blooms and buzzing hives they enhance the beauty of nature.

10. Conservation and Awareness: The situation of bees has raised public awareness about the significance of pollinators as well as the need to take conservation measures. Increased awareness can help in protecting the environment and other species that are endangered.

In short Bees are a pillar of the ecosystem with a broad range of impacts. The importance of their pollination affects economy, ecosystems as well as the wellbeing of humans which makes their conservation essential to a healthy and sustainable planet.

Benefits of keeping bees

Beekeeping offers numerous benefits that extend beyond the basic collecting of honey. These are the main benefits of keeping bees:

1. Honey Production: Producing a supply of natural, delicious honey is among its best advantages. Beekeepers have the ability to keep a steady amount of honey from their farms which they are able to buy or sell to local customers.

2. Pollination boost: If you are a gardener or cultivating agricultural crop, having a beehive nearby will greatly increase the amount of the pollination. This leads to higher production and better quality food.

3. The Nature Connection Beekeeping is unique way to connect with the natural world. The program invites you to watch and interact with bees in so that you can gain a deeper appreciation of their behavior and the complex functioning that they have in their colonies.

4. Educational Opportunities: Beekeeping offers the opportunity students to experience hands-on education. It's a fantastic method to inform your family and friends on the importance of pollinators along with biodiversity and the environmental.

5. Participation in the Community: Beekeeping is a great way to assist in building communities. Beekeeping groups or workshops let you meet fellow beekeepers and share your experience and gain knowledge from each other.

6. Reduce Stress: Caring for the health of your hives, and taking a look at the bees can be a relaxing and tranquil activity. The regular hum

of the hive and the effort required to maintain it will help ease anxiety.

7. Homegrown products Apart from honey, beekeeping also produces various other beneficial goods. Cosmetics, candles, and even skincare items can created from beeswax. Royal jelly and propolis could be able to provide health and cosmetic advantages.

8. The environmental impact: By helping pollinators, it helps ensure that local ecosystems are healthy. This improves diversity of plants along with wildlife habitats as well as the balance of our environment overall.

9. Sustainable living: The practices of beekeeping have been proven to be compatible with sustainable living. Your work directly contributes in protecting the natural environment as well as helping local farmers.

10. Earnings: Beekeeping may become a modest commercial business. In addition, you

can earn money by selling products made from beeswax, honey, as well as bee-related tasks.

11. Individual Development: Beekeeping encourages accountability, perseverance, and a keen eye for detail. The challenges and rewards encourage personal growth and an appreciation of achievement.

12. Conservation: By protecting honeybee colonies you are contributing to the conservation of this crucial pollinating species. Your contribution can aid in helping honeybee colonies thrive while they are battling threats.

Beekeeping, in essence, is a rewarding hobby that offers practical benefits as well as satisfaction for the individual and environmental responsibility. If you're looking to create honey, increase pollination or simply be in touch to the natural world, beekeeping provides many advantages that will improve your life in many of ways.

Understanding Bee Behaviour

Knowing how bees behave is crucial in ensuring effective beekeeping as well as understanding the intricacies of working the honeybee hive. These are the most important aspects of behavior in bees:

1. Behaviours of Foraging: Bees perform many responsibilities in the colony And foraging is among of them. Workers collect nectar pollen, water and propolis. They relay the location of their food sources through their famous "waggle dance."

2. Waggle Communication: The dance is an intriguing behaviour in which bees communicates its direction and the distance of a food source. The duration and angle of the dance transmit important information to the other bees.

3. Hive Defense Bees have a fierce defense of their Hives. They guard the entry point and utilize their bodies to block the entry of intruders. In the event that the hive becomes

threatened and they are threatened, they will sting it to protect the hive.

4. Swarming: It is a natural phenomenon in which colonies split to create an entirely new beehive. The queen who was the first to emerge and a portion of the employees leave for a brand new colony. The existing hive creates the new queen. Things like hive congested and availability of resources affect the behavior of swarms.

5. Temperature regulation Bees are great thermoregulators. They help keep hive temperatures steady by fluttering their wings, cooling them down, or by crowding to create heat in the colder months.

6. Royal Court Queen Bee: The queen plays an significant roles within the colony. It lays eggs and releases hormones that control the behavior of the colony and maintain harmony. Queen bees are also able to fly on the mating flight to be able to join drones of different colonies.

Chapter 2: Beehive-Related Types

Different types of beehives

There are a variety of beehives utilized for beekeeping. Each has distinctive positives and advantages. Below are some examples:

1. Langstroth Hive: It is among the most sought-after honeycomb designs. It's comprised of boxes that are stacked and have removable frames which are used to hold the honeycomb. The Langstroth Hive is very popular for beekeepers since it permits the easy manipulation and inspection of the frames.

2. Top-Bar Hive: With this design the bars are positioned on upper part of the hive and the bees build their combs downward. Hives with top bars are often favored by those who like more natural comb shapes with less interference.

3. Warre Hive also called"the "vertical top-bar hive" the Warre Hive was designed to mimic a natural honeybee's nest. It is equipped with

top bars similar to the top-bar hive, but it is placed vertically, much like the Langstroth. The Warre is a hive that prioritizes quiet as well as administration.

4. Horizontal Hive: Horizontal hive are arranged horizontally like cabinets. They may be constructed using frames, bars or removable frames and offer a unique method of beekeeping, allowing inspection through the sides.

5. Flow Hive: The Flow Hive style is meant to ease the process of extracting honey. The frames are made of plastic comb cells which can be divided to allow honey to flow when it is turned. This eliminates the need to use traditional methods of extraction.

6. Skep Hive: Skeps were originally woven baskets, which were later used for beehives. They're now less popular because of the management system and inspection limitations, however they do have a traditional rustic appearance.

7. Long Hive: Hives that are long similar to horizontal hives give more space to bees. They're popular because of their versatility and accessibility. access. They are constructed using frames that can be removed or bars.

8. Watch Hive: These hives are intended to be used for teaching purpose and permit people to watch the behavior of honeybees through glass window. The hives that are observed tend to be small and they do not make honey.

9. The sun hive is an iconic European design that is made from organic materials like straw and the reed. It is popular because of its attractive appearance and natural airflow, but it is a complex subject that requires an understanding of.

10. Kenyan Top-Bar Hive Kenyan design, just like the top-bar horizontal hive originated in Africa. It is a low-cost construction method and is often used when there is a shortage of resources.

Each style of hive has pros and drawbacks. Your selection is based on your goals in beekeeping, preferences as well as the local climate. Before choosing a hive think about your previous experience with beekeeping and management style, the surroundings, resources available, and the demands that your bees have.

Beehive components

Beehives are made up of many parts working in tandem to create a secure environment for the bee colony to prosper. Here are the most important elements of a typical beehive

1. Bottom Board: It is the foundation for the hive that provides support to the whole structure. It acts as the entrance to the hive and assists in the control of airflow.

2. Brood Box/Deep Super Brood Box/Deep Super is the place where the queen bee places her eggs, while workers take care of the brood that is developing. There are frames in which bees construct combs to

create broods, and to store pollen and honey to be used immediately by the colonies.

3. Frames: Frames are removable structures which are located within the brood boxes and supers. The frames are where honeybees make comb that is removed to inspect or harvesting honey, as well as managing.

4. Foundation: The majority of frames are built on an foundation. It is made of a small piece of wax or plastic used as a compass for the bees to construct their combs in an organized way.

5. Queen Excluder is a specialized grid which is positioned between the brood boxes as well as the supers for honey. This allows worker bees to move through, while also preventing the queen in its larger size from depositing eggs into the storage area.

6. Honey Supers are boxes added to the brood box that are set above the brood boxes to store honey that has been stored by bees.

Supers contain frames, or foundations for the bees to build combs and to fill with honey.

7. Inside Cover: The interior cover, located over the top cover, provides an extra layer of insulation and airflow. It helps in the control of humidity and temperature within the hive.

8. Outside Cover/Telescoping Cover exterior cover of the hive is its most important component. It protects the bees from weather and also provides insulation. Telescoping cover-ups often have an extended edge, which helps keep water out.

9. Entrance Reducer: A small tool that can alter the size of the entry point to the hive. This is particularly important in the most vulnerable times for hives for instance, winter. a less crowded entrance can help to defend the colony against threats and also conserve the heat.

10. Feeders: A feeder that is not required, but beneficial it can be installed inside the hive for providing the bees with food supplements

particularly when forage from nature is in short supply.

11. Varroa Screen / Bottom Board Insert This insert is used to keep track of and manage Varroa mites. They are a frequent bug that may cause harm to bee colonies.

They create a functional and sustaining space for bees in which to thrive as they procreate and store honey. The arrangement and usage of these components can differ according to the design of the hive and the goals of the beekeeper.

Affirming a site that is suitable

The place you put the beehive's location is crucial to the overall health and wellbeing of your colony. There are a few important factors to take into consideration when choosing the best location for your beehive:

1. Sunlight: Bees thrive in places that receive ample sunlight. Select a place that is able to receive minimum 6-8 hours of direct sunshine per day. It keeps the beehives dry and warm essential for the health of the honeybees.

2. The protection of the sun is essential, it's important to choose an area that offers shielding from the high winds. By placing the hive in an area that is wind-proof like against a structure, or even a natural windbreak, will help keep the hive from getting too cold or being exposed.

3. Proper Airflow: Proper airflow is crucial to avoid humidity buildup in the hive. This could cause mold or bee health issues. Beware of putting the hive in places with inadequate vents or air stagnant.

4. The source of water: Bees need water in order to cool or hydrate and then dilute the

honey they store. Pick a location in close proximity to a reliable, clean water source like streams, a pond or birdbath.

5. Accessibility Maintenance, regular inspections and collection of honey requires easy access to the honey hive. Be sure to have enough space around your hive in order for you to be able to work without disruption of the plants and structures around it.

6. Your neighbors' Comfort and Safety: Take into consideration the comfort of your neighbors and security. To prevent accidental bee interaction ensure that the hive is far enough away from the property's boundaries, fences and walkways.

7. The hive should be elevated just a couple of feet higher than the surface level will protect the hive from ants and other pests as well as improve the flow of air. But, placing the hive excessively high could make inspections a challenge.

8. Legal Restrictions: Make sure to check the local rules for beekeeping as well as the zoning regulations. There are some areas that have rules regarding hive location and distance from boundaries of property as well as the amount of hives allowed.

9. Accessibility to Forage: Find areas with an array of blooming plants that will provide your honeybees with a consistent supply of pollen and nectar. Beware of hives located near agriculture fields, which could contain pesticides that can be harmful for bees.

10. Do not allow direct human contact Although bees tend to be not aggressive, it is best to be sure to keep them far from areas with large human traffic, to reduce the risk of accidental bee stings.

11. Review Your Environment: Check the surroundings for potential dangers like predators, creatures that could cause disturbance to the bees, as well as the possibility of polluting sources.

12. Expanding Future: If are planning to expand your existing hive setup in the near future, you must make sure you have enough space for additional equipment or hives.

Making the effort to select the right location will help ensure the health of your colony as well as its longevity, productivity, and health and also promotes harmonious harmony with the surrounding.

Chapter 3: Beekeeping Equipment

The most important tools for novices

For a novice beekeeper using the proper equipment is essential for efficiently and securely maintaining your beehives. Below is a checklist of essential tools you should have within your beekeeping tools:

1. Bee Suit: When checking or working with the hives, wearing a complete beekeeping outfit, which includes the veil, jacket, and gloves, shields yourself from the stings of bees.

2. The Hive Tool is a tool allows you to break the components of hives apart, take out frames and scrape away any excess propolis or even beeswax.

3. Smoker: In hive inspections smoking provides an icy smoke that calms the bees. This makes them less aggressive, which allows the bees to move more easily within the honeybee hive.

4. Bee Brush: This bristle that is soft and gentle can be used to gently remove bees from frames as well as hive components, but without harming the hive components or frames.

5. Feeders for Hive: They can serve as a supplement to the bee diets, in particular in situations where natural forage is not plentiful or when colonies are newly established.

6. Frame Grips: These are grips that allow you to lift and move frames with confidence without harming bees.

7. Queen Catcher: A specially designed instrument used to securely trap and remove the queen for examination or manipulation of the honeybee.

8. Tools or Uncapping Knife Prior to extracting honey, using an uncapping tool, or a knife, will remove the wax caps from the honeycomb.

9. Honey Extractor: The honey extractor will make honey spin out of honeycomb when you want to harvest honey. There are manual as well as electronic options.

10. Beekeeping Journal: Keeping track of bee inspections at hives, hive behavior observation, as well as the changes in hive conditions is essential in order to make informed choices.

11. Sugar Water Spray Bottle The sugar water spray soothes bees, and makes them disperse when they are inspected.

12. Marker or Paint Pen The use of a distinct color for marking the queen bee will allow the queen bee to be identified quickly when you inspect the hive.

13. Close-toe Shoes: Wearing closed-toe shoes will protect your feet when performing work around hives.

14. The First Aid Kit having essential first aid tools available in the event of tiny accidents or bee stings is vital.

15. An extensive beekeeping guidebook or guidebook contains essential information regarding hive health in addition to bee behaviors and solving problems.

If you are working with bees, remember that security is of paramount significance. In order to avoid being stinged you must make use of protective equipment, including a suit for bees and gloves. Once you've gained the experience you will realize alternatives that better fit the style of beekeeping you prefer and your needs.

Wearing safety gear

To safeguard your skin from being stung by bees and other allergens in beekeeping clothing and safety gear are vital. Safety equipment and clothing you must consider can be found below:

1. Bee Suit, or a Jacket that has a Veil: Total protection is offered by a suit for beekeeping or a coat with a veil attached. If you want to feel comfortable while performing work in the

hive select a garment made of a lightweight and breathable fabric.

2. Wear gloves that offer good dexterity as well as protect your hands from getting stinging. Some prefer Nitrile gloves that offer more control some prefer leather gloves.

3. Boots: Wear closed-toe footwear that can reach beyond your ankles to protect from the stings of bees. Make sure they are comfortable so that they are able to stand for long periods of time.

4. Wear a hat or a helmet. Wear an hat with a brim or helmet designed for beekeeping. It will securely ensure that your veil stays in place. It prevents bees from coming near towards your face.

5. Veil: A veil is a mesh piece of clothing that allows you to clearly see while shielding your face neck from the stings of bees. Be sure that it's securely attached to your hat dress.

6. Beekeeping Gaiters: They provide extra protection, by covering your lower leg and

ankles. If you're not wearing the full suit of beekeeping They can prove extremely beneficial.

7. Smoker: Smokers are the most important piece of equipment to ensure safety regardless of whether it's placed on your body. In hive inspections and inspections, it keeps bees at ease and helps them be less prone to bite.

8. Light-colored Clothing: Select lighter clothing so as to not irritate the bees that are attracted to dark hues. Bees can be kept calm by wearing white clothing or any other light color.

9. Tuck-In: To prevent insects from entering the garments you wear, put your clothing into boots and gloves.

10. Avoid perfumes and other products with scents that could draw bees since bees are very sensitive to scents that are strong.

11. Allergy Kits: If you know that you're intolerant to bee stings it's a good idea to

bring an allergy kit (such like an EpiPen) and inform those close to you know of the condition.

12. Implement The most effective defense against bee stings, regardless of your protective gear the correct techniques for beekeeping with a professional and calm manner. Do not make jerky motions as well as loud sounding vibrating to stop bees from protecting them.

Keep in mind that a few individuals may have an allergy to stings from bees and they may be uncomfortable. Be sure to wear the right protective clothing and tools is crucial for the safety of your family and yourself when dealing with bees. While you've taken all important precautions, security should always be the first priority. Be prepared to deal with stings that could occur.

Maintenance Equipment for Hive

Tools for managing your hive are essential to maintain the wellbeing of your bee colonies and beehives. Below is a brief list of hive management tools that you'll need:

1. Hive Tool Multi-purpose tool to lift frames, taking out hive pieces as well as scraping off excess beeswax and propolis.

2. Smoker: When hive inspections are conducted the smoker emits cool, refreshing smoke that soothes bees and make the bees less hostile and much easier to manage.

3. A soft-bristled, soft-bristled or "bee brush" can be used to eliminate the bees from frames and parts of hives but without harming the bees.

4. Frame Grips and Holders The use of these devices makes the lifting and manipulation of frames easier during inspections through a solid grip.

5. Queen Catcher: A specially designed device for locking and capturing the queen in a secure manner so that she can be examined, or utilized to control the beehive.

6. Feeders: A variety of feeders are on hand to offer your honeybees more food particularly when nectar is in short supply or when they are starting new colonies.

7. Honey Extractor: The honey extractor can be used to extract honey from the comb. This improves the efficiency of extraction.

8. The removal of wax cappers from honeycomb before extraction can be accomplished with the uncapping knife or fork.

9. Strainer or Sieve: Prior to being packaged, honey is usually strained immediately after extraction in order to eliminate any debris as well as wax particles.

10. Make a detailed journal of all examinations, observations as well as hive-related conditions, by keeping a notebook or journal to keep track of your beekeeping. This allows you to observe the growth of your colonies, and take intelligent decision.

11. Spray Bottle: Spray bottles can be used to control bees as they gently carry their inspections, after it has been full of a dilute sugar solution with water.

12. Paint Pen or Marker A queen bee's body must be marked using specific colors in order to be more easily seen when the hive's inspections.

13. Container or Bucket: When conducting inspections, having containers or buckets available makes it easy to take frames, honeycombs or other objects while avoiding exposing them to surface.

14. Wire embedders, sometimes called an embedding tool can be used to insert wires in the foundation of frames so as to strengthen the comb and prevent it from sliding.

15. Brace Comb Cutter: A device designed to cut away the excess or unnecessary comb which bees might create between frames.

Chapter 4: Obtaining Bees

Colonies of bees to buy

Becoming a buyer of bee colonies, commonly called nucleus colonies (also known as "nucs," is a popular way for new beekeepers to start their career. The following is everything you need to know about buying bee colonies

1. Look for reputable local beekeepers Beekeeping organizations, or apiaries who sell nucleus colonies. A purchase from a trusted vendor will ensure that you get healthful, well-managed bees.

2. The Nucleus colonies generally available in early spring and early summer months as bee numbers increase and colonies are more likely flourish and grow.

3. Packages and. Nuc: When you purchase bee colonies, you may select a bundle of bees (a group of bees with no brood, comb or brood) or the nucleus colony (a tiny established colony that has brood, frames of bees, and a queen that is laying). Nucs are

typically recommended for novices as they offer the most solid foundation.

4. Nucleus: A typical nucleus colony has many brood frames, bees, a queen who lays and the storage of food (pollen as well as honey). Frames are numbered from 4 to 5 depending on the type of nuc. A conventional nuc usually has between 4 and five frames.

5. Health and inspection: If you can check the product's packaging before buying. Be sure to look for indicators of illness, pests as well as overall well-being. Bees that are active should be able to show an abundance of worker bees as well as evidence that broods are developing.

6. Local Adaptation: Buy bees that are used to the local area and the conditions as often as feasible. Bees that are locally bred are better at home in the local conditions and are more likely to have a higher likelihood of being successful.

7. Queen quality: A robust Queen that is healthy and strong vital for the health of a colony. If you are evaluating the colony, be sure to look for queens that are well laid and evidence of cells from the queen, which might indicate the possibility of a queen replacement.

8. transfer: Make arrangements to secure the transfer of bees from their new homes. Utilize proper ventilation, and ensure that the bees safe throughout the transport.

9. Beekeeping Know-how: It's an ideal idea to be aware of an awareness of beekeeping techniques, hive maintenance and health of bees prior to buying bee colonies. This can help you with taking care of the new colonies you purchase.

10. Make sure you are aware of any local limitations or permits required to keep bees in your region. Examine whether there are any colony or beekeeping restriction.

11. Assistance and Mentorship: Search for a mentor, or sign up to an association for beekeeping in your area. Being able to get help from experienced beekeepers who can guide you is a great benefit particularly if you're an aspiring newbie.

Purchase of colonies of bees is the most exciting part of your journey to beekeeping. For a seamless and smooth transfer for your newly acquired bees using a great deal of planning and thought.

Capturing swarms

Swarm capture is an enjoyable experience for beekeepers, and also is a great way to establish the latest colonies of bees. Below is a step-bystep guide to the capture and management of swarms

1. The identification of swarming occurs most prevalent in the spring and summer months, where bee populations are growing. You can look for signs of swarming for instance, bees congregating close to the entrance of a hive

or at its outer edges or an abrupt reduction in the activity of bees within an entire colony.

2. Gather Equipment: Take all the equipment you require like an hive box that has frames, foundations and frames as well as a bottom board as well as a cover. Verify that the hive well-maintained and in good condition.

3. Select the Best Time to Catch a Swarm The daytime in the time when honeybees are busy feeding, is the ideal moment to capture an Swarm. Swarms are more likely to fly during rain or in high winds.

4. Protective Equipment: Keep yourself safe from stings caused by bees with a beekeeping costume with gloves and veil. Bees that swarm are not as defensive generally, however precautions are still required.

5. Discover the Swarm Find the site of the Swarm. It might be on an limb of a tree, a fence, or another construction. Verify that the swarm can be accessible and secure.

6. Set the Hive up: Directly under the swarm, put the hive's box that you have prepared. The hive must have foundation-filled frames so that they can provide the bees with a place for the comb.

7. Gentle Shake or Brush: Gently shake or gently brush the bees in the hive box from the surface or branch. Certain beekeepers opt to employ a gentle brush to entice bees to enter the Hive.

8. Collect Stragglers: Once all bees are in the hive, you can gently collect the stragglers left in the branches or on the surface. For gentle encouragement to move into the hive employ a beebrush or other soft tool.

9. Make sure the Hive is secure: After all bees have walked into the hive lid in the middle and be sure the hive remains tightly closed and secured.

10. Transportation and Settling: Transfer the swarm that has been collected to the final location. Bees settle and start to build combs

and will require an easy source of food (sugar sugar or even honey) during the initial days.

11. Examine and observe the Swarm regularly to check if it's building brood, comb and storage. If needed, additional nutrition should be given in particular if money is limited.

12. Queen Check: Make thorough checks after couple of weeks to ensure that the swarm that you collect is healthy and active queen. Inspect the swarm for eggs, larvae, as well as evidence of a brood that is well laid.

Incorporating a swarm can be rewarding as well as a great way to increase the number of bees within your apiary. Take care to handle the bees with respect and make use of the appropriate equipment and give the swarm that you have collected all the food they need to thrive in the new environment they have created.

Inserting bees in the beehive

When it comes to beekeeping, putting the latest bees or an encapsulated bee swarm in a hive can be an amazing moment. This is a step-by-step guide to the installation of bees into an hive

1. Make sure that your hive is neat, well-built and is in the right location. Make sure that the frames within the hive are supported by a foundation or a comb that has been drawn for bees to construct upon.

2. Wear protective Gear Be safe from stings caused by bees with a suit for beekeeping as well as gloves and a cover-up during the process of installation.

3. Make sure you've prepared the Bees If you're planning to install a swarm of bees, remove the syrup bottle as well as the queen's cage. If you're capturing a swarm Make sure that your bees are prepared for the installation.

4. The Queen Cage is to be hung If you have an enclosure for queens, be sure to place it in between the frames inside the hive. Then, hang it on the screen one side to the bees.

Before you release the queen, bees need to initially get comfortable with the queen's scents.

5. If you're installing a set of bees the box onto the ground, allowing the bees' to settle at the lowest. The cover should be removed from the container and shake or pour the honey into the hive. Try to aim at the area surrounding Queen cage.

6. Capture Stragglers: A few bees might remain attached to the container or flitting through the air after having been shaken or pour. Rub or shake them around in the hive with care.

7. Set the frames Then, gently push each frame to the back in the hive being careful not to squash the bees. Verify to ensure that the frames are properly in their place and are evenly spaced.

8. Feeding: Give a source of food to the bees for their food, like sugar water, honey or even

sugar, especially in the early days. It helps them settle into the new hive.

9. Shut the Hive down: Set the outer and inner cover on the hive ensure that the hive is secure and aligned properly.

10. Monitor and check after a couple of days, look at the hive to ensure whether the bees have embraced the queen as their queen and have begun to build the comb. Find evidence of brood production as well as pollen collection and honey storage.

11. Eliminate the Queen's Cage Check the hive every 1 week to ensure that the bees were able to free the queen of the cage. If the queen is in the open, it is possible to remove the cage.

Chapter 5: Hive Management

Assessing the health of beehive

To ensure the wellbeing and growth of your bee colonies checking the health of your hives is a crucial aspect of keeping bees. Inspecting your hives on a regular basis allows you to observe the health of your bees and behavior and pinpoint any potential issues. This is a step-by-step guide to checking a hive

1. Select a sunny and warm day for your visit. If the weather is pleasant the bees are more active and less aggressive

2. Make sure you wear protective Gear Make sure you are protected from the stings of bees by wearing your beekeeping suits as well as gloves and a cover. While inspecting the hive security should be the top priority.

3. Smoker: Make sure you light the smoker. When you are working, apply chilled smoke to soothe the bees.

4. Smoke the Entry: Blowing an ember at the entry point of the hive in order to stimulate

the bees to move deeper into the hive, and reduce the defensive behaviour.

5. Then, gently open the hive: Take off the inner and outer covers taking attention. Carefully pry open any sealed with propolis parts using your hive tool.

6. Examine the Frames: Start by looking at the frames inside the brood or top box. chamber. Be on the lookout for

Brood Pattern Check for uniformity in egg dispersion, consistency of the pattern, as well as healthy larvae within the pattern of brood. Uneven brood patterns could indicate the presence of a problem.

Seek out the queen or other indicators of her presence, like eggs or larvae.

Worker Behavior Take note of how worker bees conduct themselves. Abrupt behaviour, dying or dead bees, and extreme hostility to the queen are indications that something is not right.

Pollen and Honey Stores: Find out the amount of honey and pollen that is kept. Bees need a large amount of food in order to keep their colony alive.

7. Frame Manipulation: Gentlely lift the frames to inspect them and be careful to not crush bees, or break the comb. Place the frames in a neat and secure way.

8. Queen Check: If you couldn't locate the queen in the frame inspection, search for her in other frames with care. Take your time and be thorough.

9. The Body of the Hive: In case you own more than one hive Repeat the frame inspection procedure for each box, starting with the highest point.

10. Supplemental Feeding: If find that your food sources are getting scarce, consider providing supplemental nutrition, especially in times of a shortage of nectar.

11. Verola Mite Test: Employ suitable methods, like the use of a sticky paper or an

alcohol wash to check the levels of mites and then take steps if required.

12. Comb Health: Check the comb for signs of illness, mold or contamination. Get rid of any combs damaged or unfit for use.

13. Make a note of your observations, findings as well as actions you take in your beekeeping diary. The journal is beneficial to track the development of your hive, and also noticing patterns throughout the course of.

14. Shut the Hive down: When the test is completed then reinstall the frames covers, and the inner cover with care. Make sure that all is fixed and aligned.

15. Watch Activity: When you shut the hive, observe the bees' behaviour in order to make sure they return to regular behavior.

The regular inspections of hives, conducted every two weeks throughout the busy season, allow you to identify and fix issues that arise, as well as ensure the health of your colony, and take an informed decision for a successful beekeeping.

The distinction between the healthy as well as unhealthy Colonies

It is essential for success in beekeeping to be able to differentiate between healthy and unhealthy colonies. Regular inspections of the hive and careful surveillance can help you in determining the health status of your colonies. This is how you can tell the distinction between colonies that are healthy and unhealthy:

Colonies are in good Health:

1. Brood Design The healthy colony should have an equally distributed number of eggs, larvae and the brood that is capped across each frame. Size and color of the brood need to be the same.

2. Queen's Presence and Activity Queen Presence and Activity: or proof of her activities (eggs and larvae) is an excellent indicator. Queens that lay eggs indicate that

the colony has the capacity of reproducing, and is flourishing.

3. The worker bee population healthy colony will have a high quantity of bees that work. The majority of them are busy as they move around the frames as well as feeding outside the honeybee hive.

4. Storage of Honey and Pollen: Bees need enough pollen as well as honey storage. Pollen is essential for the development of broods and honey is a source of source of food.

5. Relaxed Behavior: When observing healthy colonies display generally calm behaviour. Bees can be heard buzzing around the area, but they do not get defensive or aggressive.

6. Frame Condition: A healthy brood and honeycombs ought to over frames. Clean, well-constructed honeycomb is a sign of a flourishing colony.

7. Healthy Brood cells: Brood cells need to be capped correctly without any discolouration, abnormalities or sunken caps.

8. Varroa Mite Control Varroa mites exist in small amounts, for example, mites tangled with adult bees and drone cells.

Colonies that have a bad health:

1. The colony of a sick one may display an irregular or uneven brood pattern with gapping, cells that aren't capped, or even dead brood. It could be a sign of sickness, or an issue regarding the queen.

2. Queen Absence or Poor performance Queen's Absence or Poor Performance be located or there's no brood activity, it could indicate an ineffective or missing queen.

3. Population Decline: The decline in amount of workers bees could indicate an illness, lack of supplies or queen issues.

4. Protective and aggressive behavior in checks could indicate that there is an over-stressed colony or ill.

5. Deformed Brood Cells: The deformity of brood cells, like American Foulbrood could indicate illness.

6. Unsatisfactory Stored Food The colony may lack enough pollen or honey, it could have a hard time feeding itself and the brood.

7. Varroa Mite Infestation Visible Varroa Mite-related symptoms like malformed wing viruses or mites on bees that are adult can indicate the presence of a mite problem that has not been treated.

8. Signs of Disease: Check for signs of illness like chalkbrood, foulbrood or nosema. The signs can be seen from the hive with bizarre colors, textures, or smells.

Regularly monitoring and early treatment are essential for healthy colony upkeep. If you are concerned about a health issue or notice signs of an unhealthy colony seek advice from knowledgeable beekeepers and local resources in identifying and solving the issue.

Controlling Pest and Disease issues

The monitoring of challenges to pests and diseases is vital to keeping your colonies of bees active and healthy. Here are some suggestions for effective monitoring and managing pest and disease issues in your beehives:

1. Regular inspections: Check the hive on a frequent schedule to observe the overall health of the bees as well as their behavior. This will help you identify any diseases or pests earlier.

2. Varroa Mites Varroa destructor mites are known and harmful insect. The levels of mites can be tracked by using adhesive boards, alcohol washes or sugar roll that has been powdered. Take into consideration treatment options when you see levels that are above the threshold.

3. Hive Hygiene: Be aware of the cleanliness of your hive. Dead bees, debris, and a large amount of propolis could all be signs of trouble.

4. Pollen Patties: Utilize the pollen patties to supplement your protein source for the health of bees and help in overcoming illness.

5. Integrative Pest Control (IPM) Make use of an IPM method to manage diseases and pests that combines methods of mechanical control, practices based on culture and biological methods, as well as chemical treatment as an last resort.

6. Queen Health: Keep track of Queen's health and efficiency. Examine evidence of queen performances, such as brood patterns as well as egg laying, frequently.

7. Brood Health: Check for signs of illness within the brood like irregular patterns, discolouration or larvae with malformed forms. Be aware of typical brood diseases like American as well as European foulbrood.

8. Watch out for any other parasites such as wax moths and small nosema, hive beetles and.

9. Make meticulous notes of inspections to the hive, observation of diseases and pests treatment methods used and the results. This aids in tracking developments and taking informed choices.

10. Learn to Stay Up current on the most common bugs and diseases that impact honeybees in your local area. Participate in seminars and courses or talk to knowledgeable beekeepers to get assistance.

11. Keep Up-to-date: Be informed in the latest research new developments, breakthroughs and the best methods for managing bees' health.

12. Get in touch with your local beekeeping association and cooperative extension offices as well as state apiarists to get advice or assistance and for information specifically tailored to your region.

13. Beekeeping Software: You might consider using beekeeping software or apps which provides disease and pest monitoring and control functions.

14. Biosecurity measures: Put Biosecurity measures in place to avoid the introduction of diseases and spread. For instance, avoid making use of second-hand equipment which hasn't been adequately cleaned.

15. Contact experts: If find a serious issue do not be afraid to seek advice from experienced beekeepers or university experts or beekeepers in your area.

Management of pests and diseases requires vigilantness and proactive actions. It is possible to maintain your colonies of bees flourishing and healthy by staying aware and informed.

The environmental aspect of beekeeping

Environment-related conditions are crucial in the field of beekeeping as they directly impact on the behavior, health and performance of the colonies of bees. These are the main environmental elements which influence the beekeeping industry:

1. Climate and weather conditions: Temperature: The hot and cold temperatures can strain bees and hinder their ability to forage, manage temperatures in the hive or rear brood.

The amount of rain is necessary for plants producing nectar to grow and ensure that bees have a steady source of food.

Frost and Freezing Frosts in late spring and autumn freezes could cause flower harm and decrease foraging possibilities.

2. Flowers: Nectar availability Bee colonies depend on flowers that produce nectar to collect the nectar and produce honey.

Pollen Sources: A variety of pollen sources supplies honeybees with a balanced diet to promote brood production and health of the colony.

3. Chemicals and pesticides Exposition to Pesticides: Inadvertent exposure to pesticides within agricultural environments can negatively impact bee health and ability to forage.

Chemical Residues: chemical residues are able to build up in pollen and honey lowing the quality of honey and other products from bees.

4. Habitat loss and land use Urbanization: The increase in urbanization could cause habitat loss as well as less foraging options for bees.

Chapter 6: Beekeeping Season

Spring Cleaning

The preparation for spring is crucial for beekeeping since it makes sure that colonies start with a healthy and robust start to the season. Here's how you can get your beehives prepared for springtime:

1. Cleaning and Sanitizing Your Equipment Start by cleaning and sanitizing the equipment you use for keeping bees including frames, bodies of hives, as well as tools. It helps keep insects and diseases away.

2. Inspect Colonies: Check each hive carefully to evaluate the health and vigor of it. Watch out for live bees, brood activity and storage for food.

3. Food Colonies' Food levels are low, you might consider providing additional food by way of syrup or sweets. Bees need lots of food sources to help with the growth of colonies and brood rearing.

4. Varroa Mite Treatment: Establish a varroa mite control strategy. Monitoring for mites, treating using approved techniques, and using strategies for integrated pest control could be all included in this group.

5. Alternate Hive Bodies In the event that you're employing multiple hive bodies take into consideration reversing the lower and top boxes' locations. The effect is to encourage bees to climb and make use of the room to raise broods.

6. Provide Pollen Patties These contain protein and help in rearing broods in situations where natural pollen sources are in short supply.

7. Queen Inspection: Ensure the colony is home to the healthiest queen. Find indicators for the queen's activity like eggs, larvae and cap brood.

8. Replace old Frames Replace the frames you have in use with brand new frames, particularly when they're damaged or

damaged. Healthy frames help in the growth of robust colonies.

9. Ventilation: To keep moisture from accumulating within the hive, inspect the ventilation of your hive and make sure that the entrances to the hive are free of.

10. Examine the components of the hive to determine if they are damaged or degraded. Replace or repair broken pieces in order to keep the hive in good condition and functioning.

11. Get ready for swarms. The likelihood of swarming increases as colonies expand in spring. Be prepared to handle Swarms by either preventing from capturing them, or by preventing them from helping create new colonies.

12. Super Installation When colonies are quickly growing, you should build supers of honey in order to let bees keep the extra honey.

13. Eliminate Dead Colonies If you notice any colonies dying over the winter months, you should clean the equipment and hives and determine the cause of the colony's loss.

14. Make sure you provide clean, fresh water Be sure that your bees have access to clean and pure water in the vicinity of your honeybee hive. This is crucial to ensure their water is hydrated and general health.

15. Pay attention to local weather conditions and forecasts for the day. Late frosts, temperature reductions or temperatures may have an impact on colony and foraging activities.

16. Be Educated: Brush up on the spring practices of beekeeping by studying books, participating in courses, and keeping close contact with professional beekeepers.

17. The early inspection of the hive: Perform early spring inspections while temperatures rise, to gauge the strength of your colony and its progress.

The mood for the whole period of beekeeping is established during the spring. Your colonies can thrive and increase their efficiency through the vital springtime months through proactive measures to assist your colonies.

The Summer Camp and the Management

Care for your summer bee colonies and control is essential for the overall health and productivity of your honeybees. Here's a suggestion to care for your bees correctly during season of summer:

1. Maintain periodic hive inspections that monitor the health of your colony, size, and behaviour. Check frames for evidence of brood or queen activity as well as any other resources that have been accumulated.

2. Make sure you're aware of the food sources Bees can be active foragers throughout the summer. However, it's important to monitor the supply of nectar and pollen. In addition, feeding them with sugar or pollen patties are advised if needed.

3. Supermanagement: In the event that your honey supers are overcrowded Try introducing additional supers to prevent overcrowding and boost the production of honey.

4. Swarm Prevention: Keep alert for preventative measures. Check that your hive has enough room for the increasing number of bees, and check it regularly to look for queen cells.

5. The control of Varroa Mites Keep in mind monitoring the levels of varroa Mites and take control measures when necessary. Mites' infestations may cause a decline in colonies and affect their health.

6. Maintain Ventilation: Be sure your hive is ventilated so that it doesn't overheat on scorching days in summer. Make sure you have enough space for the hive and make sure to use bottom boards that are screened.

7. Water Source: Make sure that your bees are able to access an uncontaminated,

reliable water source near the honeybee hive. It is essential to ensure hive cooling and hydration.

8. Bee Health: Examine for any signs of illness like malformed wing viruses or foulbrood. Also, look for others on a frequent routine. If you spot any issues you should respond immediately.

9. Watch Queen Performance: Search for a queen who is laying and look at the brooding pattern. Queen health is crucial to the productivity and growth in the colonies.

10. Gathering honey: If you're supers for honey are already filled, think about collecting extra honey. In order to avoid causing injury to the hive or upset the bees make sure you are using the correct methods for honey extraction.

11. The Queen's Replacement: In the event that there is an improvement in the queen's performance or signs of problems with the

queen Consider changing the queen's role to preserve the overall health of your colony.

12. Preventing Robbing: Be aware of behavior that resembles robbery that occurs when the bees of other colonies try to take resources. Make sure to keep open syrup or honey containers out of the way of the hives, to lessen the chance of this happening.

13. Shade: To avoid excessive heat, offer shade to the hives during extremely hot days. Shaded areas help keep the temperature of the hive in the control of.

14. Keep a Record: Always keep accurate records of all hive visits, treatments utilized in the hive, as well as overall hive health.

15. Be informed: Keep current on the local weather, potential nectar flow, as well as any regional issues that might impact the beekeeping industry.

16. Learn to Educate Yourself: Participate in Beekeeping seminars, read about books on beekeeping, and connect with beekeepers

who have experience to know more about summer-related management techniques.

A proper summer treatment and management will ensure a well-balanced overwintering experience and a well-established colony in the coming year. If you pay attention to your bees' requirements and needs, you can help them flourish during the busy period of summer.

Winter and Fall Hive Preparations

Preparing your hives for fall and winter is essential to ensure the longevity and health of your colonies of bees during the winter months when temperatures are colder. The following is an in-depth guide for getting your hives ready for fall and winter

Fall Hive Preparation for the Hive:

1. Health Assessment: Conduct thorough inspections of hives to evaluate the general health and vitality of the colony. You should ensure that there is the health of the queen, plenty of brood and lots of food.

2. Varroa Mite medication: Apply varroa mite medication to lower mite numbers before winter. The lower levels of mites help to ensure that winter bees are healthier.

3. Feeding: Be sure that colonies are stocked with enough pollen and honey for the duration of winter. If the supply of honey is out, think about making sugar or sweet cakes to provide an extra source of food.

4. Join Weak Colonies Combine weak colonies: If you're in the middle of fragile colonies that aren't likely to weather the winter by themselves Consider merging them with more robust colonies in order to increase their odds of the survival.

5. Condense the Hive Space: Condense your cluster by removing the empty honey supers, as well as consolidating frames inside the brood room. It helps in creating a more warm and comfortable the cluster.

6. Insulate: Insulate the hives as necessary to shield honeybees from freezing temperatures.

Make use of products like foam insulation, or even insulating cover.

7. Mouse Guards: Place mouse guards in order to keep rodents from coming into the hive, causing harm.

8. Ventilation: Ensure proper hive airflow by maintaining a moderately upper entrance as well as a properly ventilated lower entrance. This will prevent humidity from building up within the beehive.

9. Examine for Disease Check for signs of illness, like chalkbrood or mildew. Take care of any problems prior to the winter season is here.

10. Weatherproofing: Cover any gaps or cracks in the hive in order to prevent the infiltration of moisture and draughts.

11. Place equipment that has not been used in storage: Put unneeded equipment in a dry, secure place to shield it from the cold winter weather.

Winter Hive Preparedness:

1. In the colder months, you should be sure to check the weight of your beehives regularly. The presence of a hive that is light could signal an insufficient supply of food.

2. In the event that you find that a colony is depleted of food items, provide emergency feed by placing sugar or fondant cakes over the frames.

3. Wind Protection: Set Hives in a safe position or create windbreaks that protect them from the harsh winter wind.

4. Reducing disturbances in winter. Only open the hive in the event of absolute necessity to stop the cold air from getting in.

5. Offer better ventilation: Install an entrance that is higher or a vent shim that allows air escape and prevent cold air from getting into the beehive.

Chapter 7: Bee Behavior And Communication

Bee communication is an intricate and fascinating aspect of bee behaviour that includes numerous signals and cues in order to transmit information throughout the beehive. Bees use communication to coordinate tasks, communicate information about resources, and help keep the hive operating smoothly. Here are the ways that bees exchange information:

1. Dancing: Perhaps the most widely-known type of honeybee communication can be described as"waggle dance. "waggle dance." Honeybees dance in this manner to signal the position of their food sources, such as pollen or nectar. The timing and direction of the dance provide information regarding the distance as well as direction of food relative to the location of the sun.

2. Pheromones. Pheromones are chemicals that honeybees utilize for communication with one another. In order to maintain the

cohesion of colonies and regulate the behavior of reproduction and prevent the development of ovaries in the worker The queen bee makes the hormones pheromones.

Beacons communicate through moving their bodies in a rhythmic manner against the comb's surface. The method of communication utilized to communicate the desire for food, or to alert threats or coordinate actions such as the swarming.

4. The Bees Communicate through touching each other. Bees, for instance are able to interact through tapping their antennas in the "trophallaxis" process in which they trade their food, as well as communicate regarding what nutrition requirements of their colony.

5. Alarm Pheromones: If the bees detect a threat then it emits an alarm pheromone that warns the other bees about imminent threat. The result is that other bees get more vigilant and prepared to defend themselves against attacks.

6. Queen Substance: A queen bee emits the pheromone referred to by the name of "queen substance" which promotes cohesion in the colony and hinders workers bees from forming the ovaries. This pheromone can also indicate the presence of the queen and her health.

7. Drone Congregation Areas gather in specific areas called "drone congregation zones" to get married with female queens that are virgin. Drones are able to locate these areas by the presence of specific indicators of the environment, like the presence of landmarks or Pheromones.

8. The Orientation Flight: To gain knowledge about the area and surrounding of the hive, bees of young age take short flights of orientation outside the colony. They can then be able to navigate and communicate the location of the hive's hive to colony members.

9. The signals of temperature and humidity Bees utilize temperature and humidity in order to keep the environment of their hives.

Environment-related factors affect bee behavior and the clustering process.

10. Wing Buzzing: Bees use their wings for making buzzing sounds, which they utilize to contact other bees. Bees, as an example can buzz to signal the arrival of a brand new queen, or warn other bees of the necessity to have more space.

Bee Communication Bee Communication

Knowing how bees communicate is crucial for beekeepers as well as researchers. It is possible to gain insight into the complicated social dynamics in a colony. We can also develop new strategies to regulate and sustaining the health of bees through understanding these messages.

Hierarchies and roles in bees

A honeybee colony is characterized by clearly defined order of things and the division of labor within the bees. Every bee performs unique tasks and duties that affect the general functioning of the hive as well as its achievement. Below is an overview of the hierarchy in the hive as well as the various types of jobs that bees play:

1. Queen Bee (one for each colony) Queen Bee (one per colony) is the center of reproduction for the hive. Its primary function is the laying of eggs which helps ensure the viability that the colony. The pheromones she produces regulate behaviour and growth of the worker bee as well as ensuring the cohesion of the hive.

2. Worker Bees (thousands in a colony) They are females that do not reproduce. They perform many different tasks such as:

The Nurse Bees The nurse bees are the newest bees that work as workers and care for the brood that is developing. They provide royal jelly to the larvae. It is a distinctive fluid that promotes development.

House Bees: When they grow older, the worker bees are transformed into house bees in which they create and keep the comb, control hive temperature and humidity, as well as store food.

Foragers: Foragers are older bees that work as workers. They harvest nectar and pollen as well as water as well as propolis, from their environment. Foragers perform dance in order to signal other bees about their food sources.

Bees that guard the hive against intruders, including other pests and robber bees that seek to steal resources.

The Undertaker Bees: The They help keep the hive tidy by taking out dead bees as well as other debris.

The bees of the Scout are always looking for fresh sources of food and water or places for hives. They share their findings with the other bees through dances and Pheromones.

Builders: Workers create combs and repairs to buildings for hives, ensuring the physical stability for the colony.

They use their wings to cool the hive as well as regulate temperatures as well as transfer hormones.

3. Drone Bees (hundreds in a colony) They are male bees, whose principal purpose is to mat with female queens that are not from other colonies. They are not involved in the foraging process or any other chores of the hive. Drones are killed after mating, or get evicted out of the hive just before the winter.

The division of work in the hive is an adaptable strategy that increases efficiency and ensures colony longevity. Bees' tasks change with age, and biological characteristics shift. The complex structure and the task specificity are crucial to the success of honeybee colonies.

Bee dancing and foraging

"Bee Dance "bee dance" is a unique form of honeybee communication that they use to notify others in the hive on the whereabouts of food sources, specifically pollen and nectar.

In recognition of the unique shape of the figure eight that bees exhibit the dance, it is often known as"the "waggle dance." The dance of the bee is a captivating illustration of how bees communicate intricate information to one another. The way it is done:

1. A worker bee comes across a lucrative resource for food, such as an area of flowers that has abundant pollen or nectar and returns back to the hive and inform that the rest of the bees.

2. Waggle Dance Returning forager performs the waggle dance across the surface of the comb. Straight-line "waggle race" is then followed by a circular path within the dance. The orientation of the straight run as well as length provide details about the direction and distance of the supply of food in the hive.

Direction Direction: The angle of the of the waggle run in relation to the vertical location of the comb will reveal the direction of the source of food with respect to the sun. If the comb's run is pointed straight upwards, the source of food is facing the sun. The food source is located 45 degrees right of the sun, if it's located at 45-degree angles then on.

Distance: the length of a dance runs shows the distance from the source of food. The farther away from the source will be, the more time the dance runs. A dance that lasts one second as an example could represent a 1-kilometer distance.

3. Audience Reception: Other bees gather around the dance bee, and can discern the direction as well as distance information. This data is utilized to determine the food source and the location of it.

4. Foraging Decision: By using the data gathered during the dance, bees will decide the best time to go foraging on their own in the designated zone. The more worker bees follow the path indicated and will forage when the food available is appealing enough as well as easily available.

The dance of the bee exemplifies the ways honeybees communicate complex spatial information via the use of hormones and signals. This is a dance that allows honeybees to rapidly transmit important information regarding sources of food, which aids the colony's hunting and harvesting performance. This dance shows how smart bees have become to survive and thrive in the environment.

Honey extraction process

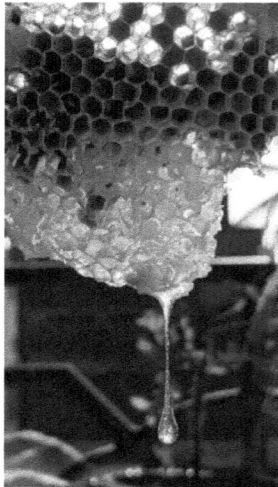

Honey extraction is the method that involves collecting and processing honey using frames of beehives to be consumed. Below is a step-bystep instruction on how to extract honey from a frames of the

Tools and Equipment

The honey extractor can be electric or manual. extractor

A tool or knife used to open a cap

Uncapping of the tray or tank

Strainer or sieve

Containers and buckets made out of food-grade substances

Honey storage containers or Jars

Appiary dress as well as gloves, a veil, and an apiary outfit.

Extraction Methods:

1. Harvest Time: Choose the time when the majority of honey frames have been capped and indicates that the honey is in the process

of being extracted. Harvest during a sunny day in which the honey flows effortlessly.

2. Install your extraction space using the honey extractor, decapping apparatus, an strainer, uncapping tank and honey storage containers.

3. Take the frames that have honey capped out of the beehive. Place them in a place which you are able to work comfortably.

4. Uncapping: Take care to remove the wax caps from each side of honeycomb using the help of an uncapping tool or knife. Place the frame on the tray or tank that is used for uncapping to collect the fallen wax caps. To speed up the process, use an electronic uncapping knife.

5. The tank for uncapping or tray collects wax caps as well as any honey drops that fall off of them. Then, you can press the cappings in order to extract additional honey.

6. Honey Extractor Settingup Place frames with no caps inside the honey extractor. The

extractor resembles a drum device that spins the frames, and then extracts honey by using centrifugal force.

7. Spin Extraction: Either manually or using an electric motor turn the extractor. Honey flies off of the frames before settling in the base of the extractor due to the rotational motion.

8. Straining: Once the honey is extracted then strain it through an abrasive screen or a strainer to get rid of any bees' remains, wax or other contaminants.

9. Bottling: Pour honey into containers that are food grade like big jars or buckets. The honey should be allowed to sit for several days in order so that air bubbles can appear on the surface.

10. Make sure you clean and sterilize honey jars using the pure, unfiltered honey after the air bubbles have increased and the honey is settled. It is easy to pour honey using an honey funnel or gate.

11. Storage and labeling: write the date, along with any other information pertinent to the container. To prevent crystallization and preserve the honey's quality, store the containers in a cool and dry location out of direct sunlight.

12. Cleaning: Following extraction, ensure that all equipment is cleaned so that you do not build up sticky remnants.

In order to maintain the high-quality and taste of honey, extraction of honey needs care and consideration to cleanliness. By following these guidelines, you will be able to allow you to enjoy the fruits of your beekeeping efforts the form of delicious fresh honey.

Propolis extraction and harvesting of beeswax

Propolis and beeswax from your hives are just two of the ways to bee-keeping and produce valuable products. Propolis is a health and medical applications. Beeswax, on the other hand, is employed for various purposes, such as the production of candles as well as cosmetics. What is the best method to collect Beeswax as well as Propolis

Beeswax Harvesting:

1. The Uncapping Process When you extract honey, you'll already be taking off the cap of your honeycomb to gain access to honey. The tank that you use to uncapping and tray using wax caps.

2. Honey Capsule Collection Following taking honey out, scrape the cappings in separate containers. It is also possible to collect the wax remnants left over from frames by using an uncapping scraper or a fork.

3. Wax Rendering: Warm the cappings collected in order to wash the beeswax of impurities and eliminate any. Place the cappings into the double boiler or solar wax melter for softening. The wax will melt and it splits away from the waste.

4. Straining: To help cleanse the wax after melting, make a strain in molds with the fine

mesh strainer or cheesecloth. It will remove all leftover debris and contaminates.

5. Cooling and storage: Allow the mold's wax to cool down and harden. After the beeswax block is been cured, take it out of the mold and store it in a cool dry place.

Propolis Harvesting:

1. Propolis Collection: Propolis is a substance with a resinous structure that is collected by bees from tree buds as well as other sources. Propolis is a therapeutic substance and helps seal gaps within the beehive. The hive tool is utilized to scrape off propolis from the surfaces of hives.

2. Scraping Method: Scrape gently propolis from the hive's surfaces, such as frames, covers, and the walls of the hive. Utilize a hive tool, or scraper for removal of the hive and not damage it.

3. Preservation: Store the propolis that you collect into a jar. Since propolis is very sticky and resinous in nature, it's suggested to keep

it in a sealed bag or container in order to prevent it from adhering to other substances.

4. Processing: To make propolis extract, or tincture make sure to soak it in alcohol with high proof (such such as vodka) in order to get its healing substances.

5. Applications: Propolis extract can be used to gain benefits for health. It is antibacterial, anti-inflammatory as well as immune-boosting properties.

Keep in mind that collecting beeswax as well as propolis must be handled responsibly and sustainable, to ensure that the health and security are not compromised. Propolis and beeswax are natural and excellent products that have numerous applications, be it in personal use or an integral part of a small-scale company.

Uses and collection of pollen

Utilizing pollen and collecting it from beehives has numerous benefits for beekeepers, and could provide positive health effects for people. Learn how to collect pollen, as well as some uses for it:

Pollen Collection:

1. Pollen traps The Pollen traps are entry devices that collect pollen when bees enter the beehive. These traps' openings are smashed against the legs of the bees, causing pollen to fall and accumulate on trays.

2. Time: Set traps for pollen in the hive at times that have a high flow of pollen that are usually in warmer weather in which flowers are abundant.

3. Pollen harvesting: Trays for collecting pollen can be taken out regularly for collecting pollen. To prevent moisture accumulation collect the pollen in a dry, cold location.

Pollen Applications: 1. Pollen is an essential food source for bees. Pollen patties, or feed containing pollen can be utilized by beekeepers to enhance bees' diet especially during times when pollen supplies are low.

2. Health Supplements: Due to the potential benefits of nutrition, certain people are using bee pollen for health supplement. Minerals, vitamins amino acids and antioxidants are plentiful in the bee's pollen.

3. Skincare and Cosmetics: Because of its potential for nourishing and rejuvenating properties bee pollen is used in skincare and cosmetics.

4. Culinary uses: Due to its distinctive taste and nutrition bee pollen can be found across many cultures for recipes as well as desserts. It is important to verify that the bee pollen is healthy to consume and isn't contaminated by pesticides, or any other harmful substances.

5. Allergy Relief: Many people consider that eating bee pollen from the local area will help ease allergies through gradually exposing your body to very small amounts of allergens. This could lead to possible reducing the your sensitivity.

6. Research and studies Bee pollen is studying for possible benefits to health, including antioxidants, anti-inflammatory and immunity-boosting properties.

If you are considering using bee pollen to eat or for another reason, be sure the pollen is collected and stored in a safe and clean way. Also, if you're contemplating having bee pollen as a supplement for possible health benefits, consult your doctor to find out whether it's suitable for you.

The importance of biodiversity

Biodiversity refers the range and variety of living forms as well as ecosystems, species as well as genetic diversity throughout the world. Biodiversity is essential for the wellbeing of our planet and to the wellbeing of all living creatures and humans, too. Below are some of the main advantages of biodiversity

1. Ecosystem stability: Biodiversity-rich ecosystems are more resistant to changes in

the environment, such as outbreaks of disease as well as climate changes and natural catastrophes. Each species has its own unique role in the function in the ecology.

Biodiversity helps maintain stability of ecosystems and reduces the risk of a catastrophic collapse of the ecosystem.

2. Ecosystem Services: Biodiversity and biodiversity ecosystems offer vital services for humans, including the pollination of crops and water purification, as well as the cycle of nutrient, management of insects as well as carbon sequestration. These ecosystem services are vital in the field of agriculture, food security in the United States, water purification, as well as climate control.

3. The Medicinal and Genetic Resources A variety of animal and plant species are rich in medicinal chemicals and genetic material that could be utilized to develop new medicines, treatments as well as biotechnological advancements.

Genetic diversity within the species guarantees resistance to diseases and also provides breeding opportunities to enhance livestock and crops.

4. Aesthetic and Cultural Value Biodiversity is a source of enrichment for cultures via the appreciation of beauty, religious significance in traditional culture, as well as recreational activities like ecotourism. Traditional and diverse practices as well as traditional solutions are intrinsically linked to the diverse ecosystems of our local area.

5. Food Security: The diversity of ecosystems offers an array of foods, which includes animals, wild plants and water resources that contribute to food security worldwide.

6. Ecological Balance: Each species is a part of an ecosystem. It helps to maintain balance and preventing the overpopulation of a specific species.

Predators maintain prey populations within control, which reduces the chance of instability in ecosystems.

7. The Biodiversity Research Institute is a rich source of scientific information, allowing researchers to study the complexity of living things and gain knowledge from the natural system.

The research into biodiversity helps us to improve the understanding of evolution process, interactions between ecosystems and the dynamics of ecosystems.

8. Climate Change Mitigation Diverse ecosystems store and trap carbon, thereby helping reduce the impacts of climate change.

Green forests, grasslands and marine habitats assist in regulating global climate trends.

9. Integral Value: Biodiversity is intrinsically valuable, this means that each species has the right for existence and to help contribute to our planet's wealth and biodiversity.

Conservation of biodiversity is a moral obligation that recognizes the value of all living creatures.

The conservation and preservation of biodiversity is essential to the long-term health of planet's ecosystems, as well as for the wellbeing of all living creatures which includes humans. This will require concerted effort at the regional, national and international levels to ensure that the next generation will benefit from the vast biodiversity of the earth.

Plants that are pollinator friendly

A pollinator-friendly garden through the cultivation of a wide variety of plants and flowers can be beneficial for the bees, and other pollinators as it improves the beauty and overall health of your gardening space. Learn how to cultivate pollinator-friendly plants

1. Choose native plants These plants are specifically adapted to the region's climate

and soil type, making these plants more attractive and beneficial for local pollinators.

2. Select a variety of flower Types: Pick the flowers that come in various sizes, shapes, and colors that will attract an array of pollinators who have varying preferences in their feeding.

3. Keep Blooming Consistently: Flowers that flower in different periods of the year in order to keep the availability of pollen and nectar through all seasons.

4. Use Pesticides Carefully: Cut down or stop using herbicides or pesticides as they could affect pollinators. Use organic and natural pest control methods.

5. Grow native plants like trees and shrubs These trees and plants provide more food sources, as well as place for the nesting of pollinators. You can plant dogwood, willow, or serviceberry.

6. Take into consideration the color and fragrance Flowers with vibrant colors often

draw greater pollinators. Flowers with purple, blue white or yellow flowers are particularly stunning.

7. Plants in Clusters: grouping like-minded plants allows pollinators to locate and gather foodstuffs. They are more attracted to larger flowering patches.

8. Incorporate Host Plants: Some pollinators like butterflies, depend on certain species of plants to host their larvae. The host plant's presence is beneficial to all stages of life.

9. Beware of Hybridized Plants: Some hybrid plants might not produce the same amount of pollen or nectar like their non-hybrid relatives. If you can, stay with the traditional varieties.

10. Offer shelter and water: Be sure to include bee structures or nesting boxes as well as the water source, such as birdbaths in order to satisfy pollinators' water and shelter needs.

11. Promote and educate: Learn about the benefits of gardening for pollinators to those

who live in your area. Help your friends, school and other community-based organizations to design pollinator-friendly gardens.

12. Keep track of and observe your garden on a frequent basis to observe the pollinators who visit your garden and what flowers they like. Adjust your garden based upon your findings.

When you plant a garden that is friendly to pollinators it is not just aiding butterflies, bees, and other pollinators but increasing the overall ecological health of your local ecosystem, and increasing the beauty of nature in your environment. The efforts you make can have an enormous positive effect on the world.

Utilizing chemicals with as little of an impact as it is

Reducing the use of chemicals of beekeepers is crucial to the health of bees, preserving ecosystems as well as the creation of healthy honey and products of bees. Below are a few steps you can do to minimize the use of chemical substances:

1. IPM: Integrated Pest Management (IPM): Make use of IPM methods to avoid disease and pests through combing strategies such as the management of hives, monitoring as well as cultural practices.

2. Promote natural methods of controlling pests for example, allowing bees eliminate dead broods, and also introduction of natural predators, such as predatory mites.

3. Resistant Bee Breeding: Choose and breed bees immune to the common pests and illnesses, which decreases the requirement for chemical treatments.

4. Regularly inspecting the hives. Check hives on a frequent schedule to detect potential problems earlier and to take preventive steps prior to the onset of pests or diseases.

5. Hygienic Behavior: Promote healthy hives through encouraging the bees and hives to

get rid of polluted or unhealthy broods, decreasing the requirement to use chemical treatment.

6. Queen Health: Be sure your queen bees are strong and healthy. Healthy colonies will be more resilient to disease and pests.

7. Healthy Nutrition: Make sure that honeybees are able to access various healthy food sources to help strengthen their immune systems and improve overall well-being.

8. The conditions and ventilation of the hive Make sure that the hive is ventilated properly and spacing in order to prevent the accumulation of moisture, which could cause illness.

9. Selective Treatment: When chemical treatments are needed, apply them in a limited amount and only when colony health is in danger.

10. Natural Hive Material: To minimize the chance of exposure to chemicals to bees,

choose raw and untreated material for frames and hives.

11. Reduction of the use of synthetic chemicals In the event that treatments are necessary choose organic or natural options that are less dangerous to bees and for the earth.

12. Learn to Stay current on the latest techniques for beekeeping in the field of pest and disease control and other treatments.

13. Monitor Varroa Mites In order to manage the population of mites, you must monitor the levels of varroa mites frequently and use non-chemical methods including powdered sugar dusting and drone brood elimination.

14. Organically Certified Practices Organic Beekeeping: If organic is the goal you are aiming for, adhere to guidelines and organic methods that have been certified.

Chapter 8: The Marvels Of Adaptation

For a full understanding of the wonder and beauty of honeybees one has to examine their anatomy. It is incredible development and adaption that is precisely designed to play their part within the ecosystem. From their top antennae, to the tips of their stingers, honeybees stand as an example of the ingenuity and creativity of nature.

On first sight it may appear easy, but a close examination will reveal a complex pattern. Their eyes are compound, allowing them to perceive ultraviolet light. This assists in the recognition of flowers and navigation. The antennae of their eyes are fitted with sensors that allow the detection of pheromones, and to communicate in a precise manner.

The most well-known feature of honeybees is their proboscis, which is a tongue-like straw which is utilized to sip nectar from flower. This crucial adaptation aids in the capturing of nectar that is then transformed into honey, the main food source for the entire colony. In addition to honey, the mouthparts of honeybees permit them to process water, pollen as well as royal jelly.

The body of the bee is split into three distinct parts: head, the thorax and abdomen. Each section is designed for certain functions, ranging from the housing of sensory organs as well as flying muscles, to holding vital organs needed to regulate respiration, digestion and reproduction.

The honeybee's anatomy was refined by eons of evolution, for a specific purpose in the beehive. The intricate web of adaptations assures that honeybees have been fine-tuned instruments perfectly designed for their crucial functions in pollination and the production of honey.

The Hive Superorganism: Roles, and Responsibility

In the hive's buzzing center the beehive is a complicated social structure which functions as a superorganism. It is an entity made up of individual bees who work in perfect harmony. This chapter focuses on the roles and responsibilities which are the basis of the operation of the hive.

In charge of the colony is queen bee. She's the mother and matriarch. The queen is the only egg-layer who creates the new generation of bees which guarantee the existence of the colony. Her presence can influence how the bees behave in the colony by emitting pheromones and dictates all aspects of reproductive activity, from peace within the hive.

The worker bees that comprise the majority of the colony, perform many duties. They act as caregivers, providing food and care for the young brood. They are foragers and seek the nectar, pollen as well as water, to maintain the honeybee hive. House bees take care of the cleanliness of the hive, tend to the queen and convert nectar into honey by a process of regurgitation and the process of evaporation.

The male bees, called drones are mainly used for that is reproduction. They are responsible for mating with female queens of other colonies, which ensures an increase in genetic diversity among the colonies. After they've accomplished their mission drones will often be removed from the hive because resources diminish during more cold winter months.

The seamless blend of these roles results in an synchronized flow of activity in the honeybee. From bustling foragers coming back full of pollen, to the queen's constant pattern of egg-laying, each bee plays a role in the health of this superorganism. As a beekeeper these functions allows you to understand the complexity of the hive and assist in its smooth function.

The Right Place to Be: Sun, Shelter, and the Surroundings

The best place to put your beehive is the initial step towards creating a flourishing bee habitat. Bees are incredibly sensitive and their surroundings play an important role in their wellbeing and productivity. In this post we'll look at the most important elements to think about when you're deciding the best place to put your beehive.

Sunlight: Bees thrive in places that receive plenty of sunshine. Try to put your beehive in a place with at least six hours of sunlight throughout the day. It is crucial to control the temperature of your hive and promoting activity in the foraging of bees.

Shelter: Although honeybees are awe-inspiring in the sun but they need to be protected from the harsh wind and weather conditions. Find a spot that has an area of shelter from the wind and fluctuating temperatures in the hive as well as reducing pressure on the bees.

Surroundings: The vegetation around your beehive affects the options for foraging available to honeybees. An array of flowers plants around will ensure a constant amount of pollen and nectar all through the season of foraging. Explore native plants of your region and think about establishing the bee-friendly gardens to provide your honeybees with plenty of food.

Accessibility: Keep your mind in mind that you'll require frequent access to the hive during checks and maintenance. Make sure that the hive is located in an area that is accessible, eliminating the necessity to walk through trees or travel through difficult terrain.

Security and Privacy bees, as with all living thing, enjoy an atmosphere of security. Place your hive in places that have a high amount of pedestrians or other disturbances in order to prevent them from disturbing the bees. The fence, the shrub or natural barriers may give additional privacy to the beehive and reduce the risk of being disturbed.

Beehive Styles: Picking the perfect hive for you

If you decide to get into the world of beekeeping, you'll come across different styles of beehives that each have their particular advantages and disadvantages. The kind of hive you select will affect the experience of beekeeping, therefore it's crucial to pick the one that matches your personal preferences and objectives.

"Langstroth" Hive It is among the most well-known hive designs with a modular structure. It permits easy expansion and manipulation of frames making inspections of the hive as well as honey harvesting simple. Langstroth is a popular hive and documented, which makes an excellent option for those who are new to the field.

Top-Bar Hive It offers a natural way of beekeeping top-bar hives have horizontal bars that bees use to construct their comb. This type of hive focuses on giving the bees more flexibility to design their own comb structures. The top-bar hives may be ideal for those who want to keep their bees in a state of the least amount of intervention and want to observe the bees' nature-based behavior.

Warre Hive: Similar to top bar Hives, Warre hives emphasize a less hands-off approach to beekeeping. They are designed to encourage upward growth of the hive similar to nature-based nesting habits of honeybees. The hives are less likely to require frequent checks and are perfect for those who prefer the least amount of disturbance.

Flow Hive It is a relatively new invention which was created in recent times, it is a relatively recent invention, Flow Hive allows for honey extraction, without the need to remove frames. The Flow Hive has a unique method that allows you to extract honey straight out of the hive using an open tap, which eliminates the necessity to employ traditional methods for extraction. Although they are convenient, Flow Hives can be expensive and may require careful control.

In the end, your choice of the hive's style is contingent on your needs, goals as well as your level of comfort with managing hives. Study each style in depth and take assistance from knowledgeable beekeepers prior to making a choice.

The Hive's Placement and the Orientation

When you've found the right place for your hive it's crucial to make sure you have the right positioning and orientation for your hive in order so that your bees have the greatest chance to succeed. Below are some other tips to take into consideration:

Hive Entrance the entrance to the hive should be facing either to the south or southeast. This position will allow the hive to enjoy the sun's warmth in the sunrise sun, while providing shelter from frigid nordic wind. The orientation also helps bees hunt early in the morning as pollen and nectar sources are plentiful.

The procedure of elevating the hive a bit off of the ground will help to prevent rainwater from getting into the hive, which reduces the likelihood of problems related to water. But be careful not to raise your Hive too high, since it could make hive inspections difficult.

Ventilation: Adequate ventilation is essential to avoid condensation of moisture inside the beehive. Be sure that there's sufficient air flow around the hive in order to keep out condensation, which could result in mold and others.

Clear Flight Path: Maintain the front area of the hive free from obstructions. Bees require a clear way to enter and leave the hive with no obstacles. Do not place the hive in proximity to trees or tall structures that might make it difficult for them to fly.

Consider Your Neighbors If you're a beekeeper within a suburban or urban area, it's crucial to take care of your neighbors. Honeybees tend to be gentle and non-aggressive, a few individuals may be concerned concerning bee stings. Be sure to inform your neighbors of your plans to keep bees, and respond to any concerns or questions that they might be having.

Monitoring and Adjustments Once you've set your hive, it is important to frequently monitor its surroundings to make sure that your chosen area continues to satisfy the demands of your bees. With the changing seasons as do the angles of the sun's and wind's patterns could change, impacting the microclimate of your hive. Prepare to make small adjustments as needed.

Wrapping up

The right place to build your beehive, and selecting the right hive design are the most important steps to take on your way to beekeeping. In providing your honeybees an environment that is conducive to their needs and a suitable shelter and shelter, you're setting the foundation for a flourishing colony. Be aware that beekeeping is an exciting and educational process, and as time passes you'll gain a thorough knowledge of how your particular place and hive's characteristics interact in conjunction with your bees' behaviour.

Chapter 9: Maintaining Your Bee Sanctuary It's A Lifetime Engagement

The creation of a good place for your bees to live is only the start of your experience as a beekeeper. When your hive has been put established and the bees are settled, it's time to begin maintaining your colony's health as well as ensuring its health as well as productivity. Below are some important points to keep in mind when you begin the journey of a lifetime:

Regular Hive Inspections Hive inspections are crucial to keeping bees. Infrequently monitoring your bees will allow you to evaluate their health, observe the brood and spot the possibility of problems. In inspections, you'll get an opportunity to watch the queen's pattern of laying as well as look for indications of pests or disease or other issues, as well as evaluate the overall health of your colony.

Seasonal Hive Management Beekeeping is a cyclical task, and the work you carry out will differ all through the time of the year. The seasons change and also do the demands of your honeybees. The months of spring and summer are with rapid growth of colonies and honey production. autumn the bees mostly hunt to find nectar and pollen, there are instances that they require extra nutrition. When there is a shortage or when you're setting up an entirely new colony, offering sugar syrup, or even a substitute for pollen could be helpful. But, it's crucial to find a compromise, since overfeeding could cause difficulties as well.

Integrated Pest Management (IPM): Pests and ailments are a challenge that beekeepers are faced with. The adoption of an integrated method for control of pests involves methods that include keeping an eye out for varroa mites employing IPM-friendly treatment methods, as well as ensuring good hygiene. If you are vigilant and active it is possible to reduce the effect of pests as well as disease on your colony.

Harvesting honey with care When it's time to pick honey, ensure that you do it mindfully and with respect to the work that your honeybees put into it. You should leave them with sufficient honey for the colony throughout the winter times. Take care to work with frames in a gentle manner during the extraction process so that you avoid damage to the combs and causing disturbance to the bees.

Continuous Learning and AdaptationBeekeeping is an evolving field which means there's always something new to be taught. Be open and curious about developments in techniques, information and advances in the field of beekeeping. Join local beekeeping groups or attend classes, and connect with other beekeepers for sharing knowledge and experiences.

"Cultivating Connections as a beekeeper you'll build a strong relationship with your honeybees. Being aware of their behaviour as well as understanding their requirements, and watching the intricate dance they perform is a great way to strengthen your connection with nature. It can be satisfying and inspiring, while reminding that you are part of the web of life that is all around our lives.

Establishing your bee's paradise will require time, energy, and love. The beehive you set up will turn into a place of relaxation and a place of delight and will also be a source of pollination as well as honey production. While you explore the different aspects of placing your hive as well as hive design selection and ongoing management of your hives keep in mind that beekeeping is an adventure of learning as well as growth and connections. The more you know about your bees as well as their demands and requirements, the more enjoyable the experience of beekeeping is going to be.

The Best Place to Live: Sun, Shelter, and the Surroundings

Finding the right site for your bee sanctuary is an essential step to the long-term success of your beehive. Bees are very sensitive to their surroundings, so providing the proper conditions will greatly affect their health and productivity. Here are the main elements to think about when deciding where to put your beehive.

SunlightThe bees are the occupants of the sun. Make sure to set your hive within an area that gets ample direct sunlight most likely for at least 6 hours per day. It helps to regulate your Hive's temperature, helps keep the humidity level in check and helps encourage the honeybees' foraging habits.

Shelter honeybees love the sun, they require shelter from the harsh conditions of weather. Think about locating the hive at an area that provides nature-based shelter from the strong winds. This can help reduce changes in temperature inside the hive as well as decrease stress on the bees in the event of a storm.

Surroundings: plants that surround your beehive play important roles in providing food the bees. The variety of flowering plants around will ensure a constant flow of pollen and nectar during the entire foraging season. Explore local plant species for a pollinator-friendly habitat which meets your honeybees' nutritional needs.

Accessibility: Bear on your toes that you'll require be able to get access to your hive on a regular basis to check it out and conduct maintenance. Pick a place that is accessible and doesn't require walking through difficult terrain. It will make the tasks of keeping bees easier and fun.

Privacy and securityBees flourish in a secure environment. Be sure to avoid placing your hives near areas of high traffic or that are frequented by people. In providing security and keeping the area free of disturbances, you will create a peaceful environment where your bees are able to flourish and not be stressed out.

Beehive Styles: Choosing the perfect hive for you

If you're looking for the styles of beehives, there are many options available Each with their own benefits and advantages. The type of hive you choose can affect your experience with beekeeping So it's crucial to select one that is compatible with your objectives and degree.

Langstroth Hive The Langstroth hive is an extensively popular and adaptable choice. It is made up of stacked boxes (supers) with frames to allow the bees to create their comb. This layout allows easy monitoring of hives, which includes honey extraction and inspections. Langstroth beehives are popular with novices due to their ease of access and documented practices.

Top-Bar Hive If you're looking for the more natural way of beekeeping, you should consider the top bar beehive. This is where bees create their combs using hanging bars. The hives with top bars are characterized by minimal interventions and permit bees to exhibit their own natural behavior. This type of hive may need lesser equipment, but it also offers a distinct perspective on the beekeeping.

Warre Hive As with the top-bar beehive The Warre Hive is designed to encourage bees to construct comb horizontally. The hive is designed to emulate nature of bees and encourages minimal interfering. Warre hives can be a great option for people who prefer an approach that is hands-off and love watching bees in their most nature-like form.

Flow Hive Modern technology The Flow Hive simplifies honey extraction. It is a revolutionary system that permits honey to be extracted directly from the beehive using conventional extraction techniques. Although it is useful, Flow Hives can be more costly and need careful supervision.

Selecting the best place for your honeybee sanctuary as well as choosing the right hive design are important decisions that influence your experience with bees. When you offer your bees an ideal environment and arrangement, you'll create the conditions for an enjoyable and satisfying life-long experience in beekeeping. Keep in mind that it is an ongoing process of learning, and your knowledge about your honeybees' habits will develop with time.

If you are embarking on a journey to beekeeping, feel proud in creating a sanctuary for these amazing creatures. The bee sanctuary you create can not only help the overall health of the ecosystem around you, but will also provide you with an insight into the fascinating life of bees. Please contact us with any questions or require further information regarding any aspect of keeping bees. Enjoy your beekeeping and let your hive flourish!

Beekeeping Basics: From Smokers to Hive Equipment

Making sure you have the best equipment is crucial to ensure successful and fun beekeeping. Tools you choose to use are not just beneficial to your security, but they will allow you to manage effectively your bee hive. Let's look into the essential items you'll require for beekeeping:

Smoker A smoker is a beekeeper's most trusted buddy. Smoke releases cool air into the hive thereby calming bees, decreasing the likelihood of them to be stinging when they inspect. The smoke that is released near the entrance to the hive and surrounding the frames will help make inspections more smooth and less disruptive to the bees.

Hive Tool The hive tool an all-purpose tool that can help to disassemble hive parts as well as separate frames and scrape the wax and propolis off. Its curving end can be utilized to lift frames, and to remove particles. The hive tool is essential for effective management of the hive.

Bee Brush: The bee's comb features soft bristles that permit users to softly remove bees from components and frames, without harming the components or frames. It's especially useful when you need to move insects away from places where you're required to look at or alter.

Feeders can be useful in giving food supplements to your honeybees in times when forage sources are not available or at the establishment of the hive. There are a variety of designs that include entrance feeders as well as top feeders. They also let you provide sugar syrup for the nutritional requirements of your honeybees.

Queen Excluder This tool can be used as a grid that stops the queen from producing eggs in certain areas of the hive. Usually, this is in honey storage supers. storage. Queen excluders can help control brood patterns and make sure that the supers of honey aren't contaminated with brood.

Entrance Reducer Entrance reducers can be used to restrict the size of a hive's entrance. They are especially useful in the establishment of a hive as well as when you need to guard a weaker colony from invaders.

By reducing the size of your entrance, you will help stop robbing by others bees.

Feeding Containers or Jars: They are utilized to provide bees with sugar syrup, or any other food items. They are especially useful for colonies that require more nutrition, and during periods of limited available forage.

Bedding the Part: Dressing for Behkeeping Clothing and Protection

Beekeeping clothes are designed to protect you from bee bites as well as allow you to move comfortably and safely around honeybees. What you'll need to look the part:

Beekeeping Clothing The full suit of beekeeping offers complete protection. Choose suits that feature Velcro cuffs that are elastic, zippered closures as well as attached veils. The suits are made of various fabrics like cotton, polyester and ventilated materials, providing diverse degree of comfort and airflow.

Veil It is essential to protect your head and face from stings by bees. The veil should be secured to your outfit and provide adequate visibility. You can choose between round veils fencing veils, or Hooded veils, depending on the style you prefer.

GlovesBeekeeping Gloves are available in a variety of designs, such as canvas, leather and Nitrile. Pick gloves that are comfortable and permit you to operate at a comfortable level. Certain beekeepers prefer working in gloves, to increase their agility, while beginners typically use gloves to boost their confidence.

Socks and Boots You should wear boots that are able to cover your ankles. Tuck your pants inside them so that bees aren't creeping all over your legs. The boots designed for Beekeeping provide additional security against the stings. Opt for lighter-colored socks in order to make it easier to spot hitchhiking bees.

Or, in other words

Making sure you have the appropriate equipment and protective gear is crucial to have a safe and enjoyable experience with beekeeping. When you use these items correctly, it improves your efficiency in safety, confidence, and efficiency while you look after your colony of bees. Keep in mind that your beekeeping clothing and tools are investments for your health and well-being as well as the wellbeing for your bees. While you're working alongside this amazing creature your tools and equipment you pick will serve as your ally in creating harmony with your beehive.

Becoming a Beekeeper in the form of packages, nucs or Swarms The acquisition of the first colony of bees is one of the best steps on your journey to beekeeping. There are many options to getting bees, each having specific concerns. Explore the three most common techniques that include packages as well as colony nucleus (nucs) as well as the swarm.

Packages: Bee packages include a set amount of bees that work and one queen that has been mated. These are typically packaged in box-types that have screens and may be an ideal option to establish an entire colony. The packages are usually available in the spring time and are popular with newbies. But, they need careful installation and a steady diet to ensure that the bees have a healthy colony.

Nucleus Colonies (Nucs): Nucs are small colonies, with frames that contain broods, bees, as well as a queen that is laying. Nucs can give you a head-start because they have queens that are functioning as well as brood patterns. This allows for faster colony development and is usually advised for those who are just beginning to beekeepers. They are readily available during spring and the beginning of summer.

Swarms: Capturing natural swarm is a thrilling method to get honeybees. Swarms are groups of honeybees that are led by a queen who leave the colony they were part of to create the new colony. They are typically seen between spring and the summer. Although capturing swarms is satisfying, they require expertise and knowledge to guarantee an efficient transfer.

Chapter 10: How To Make Your Bees Feel At Home

The day of installation is an important moment in your journey to beekeeping. This is the day that you welcome your bees into their new home and create the conditions to ensure their success and growth. This is how you can ensure that your bees feel welcome at home

Get Your Hive Ready Prior to the arrival of bees check that your hive has been correctly set up and is fully prepared. Make sure that your frames are properly set up and that your hive is clear of any debris. Set up any essential food sources like pollen patties, sugar syrup in the beehive.

Pick the right time: Put in your bees on a quiet and mild day, when the temperature is at or above 50 degrees Fahrenheit (10degC). Bees are more gentle in warm weather, which makes the process easier for both the bees.

Be gentle with them: They are more likely to remain peaceful when they're cool and full of food. The bees are lightly sprayed with sugar-water prior to installing This keeps them occupied when they are cleaning themselves. It also assists them in bonding to their new hivemates.

Queen IntroductionNOTE: If the package or nuc does not have an identified queen, it is necessary to introduce the queen into the colonies. Follow the instructions for the way you've chosen to introduce her for ensuring that the colony will accept her.

Attention and patience Once you have installed, give your honeybees time to get used to their new surroundings. It's common for them to investigate their new surroundings prior to constructing the comb as well as hunting. Check their activities and behaviors throughout the time following installation to make sure they're making the right adjustments.

The process of acquiring and establishing the first bee colony an amazing and instructive journey. Each approach has its pros and cons as well as the process of installation provides a chance to gain knowledge about the behavior of bees as well as the dynamics of hives. If you approach the task with a sense of patience, care and an innate sense of wonder and curiosity, you're setting the foundation for a colony that will benefit your beekeeping endeavor and to the overall health of the ecosystem around you.

Checking Hive Inspections: What to Check, What, When and How to Examine

The regular inspection of your hives is an essential aspect of success in beekeeping. They let you monitor the behavior and health of your colony, check the availability of your resources and to make quick adjustments.

We'll look at the most important aspects of inspections for hives:

Regularity: Hive inspections should take place around every 7-10 days during the busy beekeeping season. At first it is possible to check more often to make sure the colony is growing and establishing itself.

The weather conditions: Pick a day that has mild temperatures. This is when the bees are active foraging and are less likely to become defenseless. Do not inspect on windy or wet days as they can be stressful for bees.

Objectives of Inspections When you inspect they will be looking at the population of the colony, its brood patterns, the availability of resources (nectar pollen, nectar, and honey) along with the overall health of the hive's components.

Method: If you are performing a hive inspection begin by gently smoking at the entrance of the hive to relax the bees. Inspect the hive, and then work carefully through each frame. Check for indicators of the health of a queen (eggs or larvae as well as capped brood) as well as observe the bees' behaviour as well as check the presence of any diseases or pests.

Resource Management: Monitor the storage of honey and pollen. Make sure there's enough honey in the store to sustain the colony during winter. If the colony does not have enough resources, think about feeding the colony with supplements.

Record-keeping: Keep an hive's inspection log in order for keeping track of your observations and steps taken in every inspection. The log will allow you to keep track of the progress of your colony and help you make better decision-making in the future.

Spotting Issues: Identifying Health Problems and Pests

The ability to spot and address bugs and diseases is vital for maintaining the health of your hive. Routine inspections help you detect potential problems early and then take the appropriate actions. Here are a few common ailments and pests you should be on the lookout for:

Varroa Mites Varroa destructor poses a serious danger for honeybee colonies. Check for the presence of mites on adult honeybees as well as in brood cells. Develop an integrated pest control (IPM) program to deal with mites and their infestations.

Nosema: Nosema is a fungal illness that impacts bees' digestive systems as adults. It is characterized by dysentery in the entrance of the hive. Check for the presence of fecal staining. weak bees.

American Foulbrood (AFB): AFB is an extremely contagious bacterial infection which affects brood. Examine for punctured, sunken or discolored brood cap caps as well as an unpleasant odor. If you suspect something is wrong, call the local authorities to get advice.

European Foulbrood (EFB): EFB is yet another brood-related illness which is caused by a bacterium. Check for patterns of spotty brood or twisted larvae. Also, look for greyish hues.

ChalkbroodThe term "chalkbrood" refers to Chalkbrood can be described as a fungal disease that can affect brood development. Find chalky white bodies in cell walls.

A small Hive Beetle: This insect is found in weak colonies. Examine for adult beetles as well as larvae within the hive as well as on the frames. The larvae of beetles can cause damage to combs as well as honey storage.

Wax Moths: The larvae may damage combs and stored honey. Examine for webbing and larvae on the corners and combs of frames.

Controlling hive health as well as behavior with regular checks is vital to ensuring the health of your bees. If you're attentive, active and informed about the possibility of bugs and diseases You're making important decisions to maintain the health and strength for your colony. Be aware that keeping bees is a process that improves with time and every check-up is a chance to gain knowledge about the bees you have and what they require.

Reach out for more information or require further information on inspecting your hive, diseases or pest control as well as any other aspects of keeping bees. Enjoy your beekeeping and may your hive flourish with a buzzing and harmonious hive!

A Bountiful Harvest: When is the best time to collect Honey

The process of collecting honey is among the most enjoyable aspects of beekeeping. However, it is a delicate process that requires precision monitoring and timing. Being aware of when you should collect honey will ensure that your bees are able to store sufficient reserves of honey to use and also provide you with the delicious reward of gold in liquid form. These are signs to suggest it's the right time to collect:

Capped Frames Honey is ready for harvest when the frames are covered by wax. Capping signifies that the moisture content in the honey is sufficient to allow for storage over a long period of time.

Full Supers for Honey When you're making use of honey supers (boxes specifically designed for storage of honey) Full supers can be a sign when it's time to harvest. Honeybees frequently move their honey from lower supers into higher ones to ensure better temperature control and therefore, a full super will not significantly affect the storage capacity of their winter food.

Weight comparison: If you own a hive scale you'll be able to track fluctuations in the weight of your hive. The sudden decrease in weight might signal that the bees have consumed honey that is stored, which can indicate that it's time to gather.

Weather and Foraging patterns An abundance of nectar as well as a frenzied foraging activity is a sign that honeybees are busy collecting nectar and then converting it to honey. If you observe a regular pattern of returning bees with empty pollen baskets is a great time to examine for capped honey frames.

Techniques for Harvesting from Frames to Jars

After you've decided that you're ready to collect you must follow correct procedures to protect the purity and authenticity of honey. This is a step-by-step procedure for collecting honey:

Preparation: Collect the honey extraction tools such as an uncapping knife or fork, extractor, filtering buckets, and collecting containers. Make sure that the equipment you use is clean and well-sanitized.

Begin by removing the caps from your honey frames. Utilize an uncapping knife or a fork to take the wax caps from the cells. Uncapped cells allow the honey to escape during extraction.

Extraction: Put frames that are not capped in an extractor device that rotates the frames to remove honey using centrifugal force. Extractors are available in both manual and electronic models. Use the frames with care to prevent damaging the comb.

The honey needs to be filtered. Once it has been extracted from the jar, it is likely to include wax bits along with other small particles. You can filter the honey using an ultrafine mesh filter or cheesecloth in order to eliminate the impurities.

Bottles: pour the filtrated honey into clean and sterilized Jars. Make use of a funnel to prevent spills. Close the lids on the jars and keep them in a dry, cool location far from direct sun.

Bee Careful: Make sure that you leave enough honey to your bees ' survival throughout winter. Make sure not to harvest too much since this may reduce the strength of your colony.

Take pleasure in the fruits of your Work: The honey you harvest is a tasty reward for all your perseverance and commitment to beekeeping. Toast it up, use it in your recipes, or even as a sweetener for the beverages you love.

Honey harvesting is the result of all your work as an beekeeper and gives you the chance to enjoy the actual benefits of your bees' work. When you recognize the indications of a plentiful harvest, using the right harvesting methods and methods, you'll not only protect the wellbeing of your colonies as well as enjoy the unique pleasure of eating honey directly out of your hive.

Chapter 11: Beeswax Wonders Making And Uses

Beeswax is an organic result of the process of making honey is an incredibly versatile and highly valuable material with many possibilities. In addition to its use in candles, beeswax is an ingredient that opens up an array of craft and practical applications. This is a look at the amazing properties of beeswax

Candles: Beeswax candles give off a honey-scented warm glowing glow. They are also popular due to their clean, natural burn. Beeswax candles are usually selected for their low influence on the quality of indoor air.

Cosmetics and SkincareThe beeswax's qualities to moisturize makes it an extremely popular element in all natural makeup as well as skincare products. It aids in retaining moisture, and also creates a protective layer for skin.

Lip Balms Beeswax is an essential ingredient in lip balms. It provides nutrients and protecting qualities to ensure lips remain soft and moisturized.

Wood and Leather Conditioning Beeswax-based wood conditioners and leather aid in preserving and protecting these products from drying and fracturing.

DIY Beauty Products from salves and lotions to body butters and balms Beeswax gives a luxuriant appearance and benefits in homemade cosmetics.

Food Wraps: Beeswax-coated fabrics can be utilized to create recyclable food wraps that provide an environmentally friendly alternative for plastic wrap.

Creative Crafting Beeswax is a great material to use to create a myriad of craft projects for example, encaustic art (painting using melting beeswax and other pigments) as well as modeling and even as an adhesive.

Propolis, Pollen, and Royal Jelly: The Hidden Gems

Although honey is by far the most popular product from beekeeping, the hive produces additional valuable products that have distinct benefits.

Propolis is often called "bee glue" propolis, also known as "bee glue," is resortable substance that bees gather from plants and trees. Bees utilize it to fill fractures and crevices inside the hive as well as to keep away invaders. Propolis is antimicrobial as well as antioxidant properties, and can be found in many health and skin care products.

Pollen Collected by bees that work Pollen is an important energy source for the beehive. Beekeepers can harvest pollen to gain its nutritional value, as well as for possible health benefits. Many people use pollen for a nutritional health supplement.

Royal Jelly Royal Jelly is produced by bees on the job and exclusively fed to small larvae, which are then destined to be queens. Royal jelly is a rich source of protein, vitamins as well as minerals. It is often found in diet products and cosmetics.

Beyond the honey pot the beehive has a wealth of unique items that could improve your life in a variety of unexpected ways. From the enthralling possibilities of beeswax, to the nutrients and health positive effects of propolis, pollen as well as royal jelly every one of these products is a different story about the fascinating world of honeybees. While you are on your journey to beekeeping, you should consider studying the various uses for the amazing ingredients.

Contact us for more information or want to dive further into the realm of beehive products, making using beeswax, or other aspects of beekeeping. Have fun exploring, and we hope find the hidden treasures in your honeybee hive!

This is the Bee-Pollinator Connection: Why Bees matter

Bees play an important part in the environment as well as our food systems by pollinating of flowers, which includes most of the fruit, vegetables, and nuts that constitute the majority of our food. Bees' intricate relationship and pollination can have profound consequences to both the natural world and our own society.

Ecosystem HealthBees are a key contributor to the biodiversity and health of ecosystems through the growth of plant species. In turn, this helps support a variety of different species, such as birds to insects, and mammals.

Food Production: About three-quarters of all food production depends on the pollination through bees and other pollinators. A variety of crops, including the berries, almonds, apples and coffee are dependent on pollination for large and quality harvests.

Biodiversity: Bees contribute to maintaining the diversity of biodiversity by encouraging the diversity of genetics within populations of plants. They also help to strengthen the resilience of ecosystems against environmental change and stressors.

Economic Effect: value of the pollination benefits provided by bees is estimated to be in the hundreds of billions of dollars. The loss of populations of bees could have a wide-ranging impact on the food industry, prices for agricultural products, as well as the lives of farmers.

Cultural and aesthetic value beyond their economic value bees hold an aesthetic and cultural significance. Their hum of activity adds the beauty and vibrancy of the natural landscapes and gardens.

Sustainable Practices: Helping Bees Thrive

If you are a beekeeper, then you are able to aid in the wellbeing of the population of bees and to protect the natural environment by implementing environmentally sustainable methods. There are a few ways to aid in the growth of bee populations:

Plants with Pollinator-Friendly Flowers You can cultivate many blooming plants in your yard or area for beekeeping to provide honeybees with an extensive and constant supply of nectar and pollen.

Reduce the use of pesticides. Avoid the use of pesticides that contain chemicals or herbicides as they could harm bees, as well as the other insects that are beneficial. If you require pest control, consider alternative options that are organic and non-toxic.

Give them water: Bees require access to drinking water for hydration. Make shallow water sources like birdbaths, or water trays that permit the bees to drink in a safe manner without risking drowning.

Implement Integrative Pest Management (IPM): If you are experiencing problems with pests or diseases that impact your colony, you should employ IPM strategies that are geared towards the health of pollinators and the surrounding.

Instruct others: Discuss the information you have learned about bees and their significance to families, friends, and the neighbors. Inspire pollinator-friendly gardening practices as well as landscaping.

Be a Bee-Friendly Advocate: Demand for policies to protect the habitats of pollinators as well as their food sources. Promote initiatives that encourage the sustainable cultivation of land and restore habitats.

Beekeeping is an amazing relationship between man and nature that will have a profound positive effect for our food and environment ecosystems. When you recognize the importance of pollination from bees and implementing sustainable methods, and sharing your expertise and experience, you're contributing to the well-being and sustainability of communities and ecosystems. As a responsible steward of bees, you hold the ability to sustain this fragile connection, and to make an impact on the world.

Controlling Swarms: Prevention and Management Swarms are a naturally occurring part of the bee colony's reproduction process, however they also present problems for beekeepers. The ability to prevent and manage swarms efficiently is essential to maintain the stability and health of your honeybee colony.

Prevention:

Monitoring Colony Condition Monitoring the hive regularly allows you to gauge the strength of colonies as well as space availability as well as queen's efficiency. Find out what causes to swarming such as the overcrowding of the hive or a lack space.

Provide adequate space Make sure your hive is equipped with enough space to store pollen and honey. In addition, adding supers when necessary to stop overcrowding.

Replace old queens: The older queens tend to cause swarms. Think about requeening colonies using queens in their prime, to limit the likelihood of swarming.

Management:

Early Detection: Keep on the lookout for indications of swarm-related preparations including eggs laid by queen cells being less by the queen, as well as the increase of drones.

When splitting hives, you observe swarm preparations think about splitting. Make a new colony by combining certain bees and other resources of the hive that was originally. This could help stop the original colony from spreading.

Add Space: If the hive has a good chance of being swarming, but isn't quite that by adding supers or frames that are empty may give bees greater space and can deter them from the bees from swarming.

Catching Swarms an swarm leaves the hive, it is possible to take it to the nearest hive and bring it back to the empty hive, or to nuc. It can provide you with an extra colony, and also keep bees from being lost.

The Queen's Issues to be solved: Replacement and Supersedure

Queens are vital for the health and efficiency of the honeybee. Queen-related issues may arise from genetics, age and other causes. How to deal with the most common issues related to queens:

Queen Alternative:

The Queen is either old or in decline If you see an absence of eggs or brood pattern that is not as good or any other indicators of a weak queen is it the right time to change her.

Introduction of a Queen The process of introducing a queen is possible using several methods, including making use of a queen cage, or directly release. It is important to adhere to the correct protocols to make sure that the queen is accepted by the entire colony.

Supersedure:

Natural Supersedure: Sometimes colonies will replace its queen by natural. It could be due to old age, or some other reason. Be aware of the behavior of the colony as well as whether there are queen cells.

Queen Cells: If observe queen cells growing examine their position as well as the number. If you notice several queen cells in an image, it is possible that it could be that the colony is preparing for a swarm.

Becoming aware of common challenges in beekeeping such as problems with queens and swarming require attention, expertise as well as prompt intervention. When you're aware of the behavior of your hive and its needs and responsibilities, you will be able to effectively avoid and address these issues so that your colony can flourish and ensure a positive beekeeping experience.

Chapter 12: The Study Of Bees Breeding And Rearing

After you've established yourself as a beekeeper may be intrigued by the realm of bee breeding and raising. Breeding bees lets you choose specific traits like resistance to disease efficiency, temperament, or productivity. This is how you can get started:

Selecting Breeding Stock Choose colonies that have desirable characteristics as well as genetics. Seek out colonies that have an excellent temperament, high honey production and a the ability to resist diseases.

"Queen Rearing": This is the process of the raising of new queens using chosen breeding stock. There are a variety of methods available including grafting, for example using queen cups. Learning to rear queens requires the practice of and understanding about bee biology.

Genetic Diversity: Keeping your genetic diversity is essential to avoid inbreeding and increase the general well-being of your the apiary. Beware of using the same queen over many generations.

Notes on Recording Data Make meticulous notes of the traits you're choosing to breed for as well as the performances of various queens. These records will allow you to take informed decisions about breeding.

Expanding your Apiary Bead Collector to Urban Beekeeper

When your skills and confidence increase, you may think about the expansion of your apiary or take your beekeeping skills to the next level.

The Urban Beekeeping A lot of cities are now allowing beekeeping in urban regions. Review local rules and regulations as well as the possibility of the possibility of establishing beehives on rooftops balconies or in community gardens.

Manage multiple hives Expanding your apiary involves taking care of multiple Hives. This provides invaluable experience on hive management, diseases prevention and the dynamics of colony.

Niche Products: When you have several honeybees, you are able to experiment with products made from honey that have a niche including monofloral honey (honey that comes from a particular plant) and creamed Honey.

Learning from others: As you get more proficient, you will be able to teach new beekeepers by hosting presentations, seminars or mentorship programmes.

Continuous learning Beekeeping is a process that continues to grow. Keep up-to-date with the latest research, best practices for beekeeping as well as the latest innovations for the coming years.

Enhancing your experience in beekeeping through exploring breeding for bees and growing your apiary could open opportunities to expand your horizons as an beekeeper. This will help you increase the understanding of bees, help the beekeeping community as well as play an even bigger contribution to ensuring flourishing and healthy bee populations. Take each challenge and chance with a sense of excitement, and keep in mind that beekeeping is a process of development and learning.

Chapter 13: What Is Exactly Beekeeping?

Beekeeping, or apiculture, refers to the maintenance and management of honey bee colonies. They are usually found in the form of hives made by humans.

One of the most ancient forms in food processing is the beekeeping method, that is the method to artificially maintain honey bee colonies. Beekeeping, often referred to as apiculture is thought to have taken place at least 13,000 BC. Since honey was a crucial ingredient in their diet The ancient Egyptians excelled at keeping bees.

Though most people associate beekeeping as the production of honey, it is possible to find a variety of other ways that modern beekeepers can to earn money out of their colonies. Beeswax as an example is commonly used in manufacturing of cosmetics and candles. An extremely popular food supplement is royal jelly. It's a substance that is released by the hypopharyngeal glands in workers in their early stages. Propolis is a compound of resin created by honey bees in order to close hive cracks is used in alternative therapies such as such as acupuncture, homeopathy, and acupuncture. A lot of commercial businesses in beekeeping provide services for agricultural pollination, which are a significant part of the revenue they earn annually.

The term "bee yard" or an apiary is in which a beekeeper keeps his bees. The colony of bees is located within a hive that comprises a set of wooden frames and boxes which carry wax sheets that the bees can make use of as a base in the process of creating honeycomb. The top container is filled with honey while the lower box is home to the queen bee as well as most of the workers bees. The Langstroth bee hive can be described as the most commonly used type of hive used in keeping bees in the United States.

Since honey bees are hazardous, a beekeeper should be aware of various safety precautions while working in proximity to a honey bee colony. Hats or vests are commonly worn to shield your neck and face from being stung. They are an additional form of beekeeping safety, though some beekeepers feel that they restrict their movements. The hooded suits, typically comprised of light-colored material is also a good option for aid in distinguishing the beekeeper against honey bee's natural enemies.

Smoke is a tool used by beekeepers to aid quiet honeybees when working with a colony. Smoke is essential for beekeeping as it blocks the guard bee's pheromones and also encourages bees from other colonies to eat, by fooling people into thinking that they'll need to get rid of their hive shortly. It allows the beekeepers to inspect the colony, and to make any repairs that are required. The fuels which can be utilized for a smoker bee include wood chips, paper pulp or corrugated cardboard Also, compressed cotton.

Most people decide to keep bees due to their fascination with bees and are looking for a pastime which will help their gardens and the natural ecosystem.

Then, there are those who are looking for the delicious honey as well as other bee-related products in high demand that your bees could provide.

What exactly does a BeekEEPER Do?

The beekeeper takes part with all aspects related to growing as well as propagating and producing honeybee-related products.

The beekeeper has to keep his beehives fit and healthy.

The beekeeper should keep notes and be vigilant about the health of and the numbers of their bees in order to accomplish this.

The beekeepers must also ensure that the hives in a good state as well as harvest honey and other products from bees.

Why is it beneficial to keep a beekeeping account for HOMESTEADERS?

Homesteaders everywhere are seeking ways to be more independent, and beekeeping is a great way to aid them on their way to becoming self-sufficient.

Honey bees that are kept in the honeybee hive produce not the most delicious honey to home for consumption. They also increase the production of vegetables as well as is healthy for the environment and can bring in additional income.

The setup of a small beehive, or even two, doesn't need a huge amount of cash or space. In fact, managing the two hives one hour each week.

Take a look at beekeeping for enjoyment as well as profit, if you're gardener looking for a profitable venture.

7 different ways to earn Money by Keeping a Book

Beekeeping could be a profitable hobby that provides the natural sweetness of honey as well as aids in pollination of the plants.

The majority of backyard beekeepers who are attracted, ask, "How does honey make profits?" and "can beekeeping make money?"

Beekeeping can be lucrative, and although selling honey may be one way to earn a profit however it's not the only option.

Here are a few various ways in which beekeepers can earn money.

1. BEE PRODUCTS

One way to earn money is selling honey products they produce.

While most people are accustomed to thinking of honey, bees also produce various other items that are taken and then offered for sale.

HONEY

Raw, local-produced natural honey is in high the market, and production costs are very low.

It is possible to bottle, sell, or sell honey that has distinct flavours, if there are particular nectar sources close to the hive.

There is a higher price in natural honey than the mass-produced honey sold in grocery stores that can be as high as $10 for a kilogram (PS15 for a kilogram).

Remember that you'll most likely not have the ability to collect honey during the initial year, as your bee colony will need some time to settle prior to producing enough honey.

The quantity of honey that is that a beehive produces fluctuates each year. The reason for this is climate, the availability of nectar and the overall health of the colony.

POLLEN

Bees take pollen from the flowers and bring it back to the honeybee hive. This is their primary energy source, and they need pollen for food to the colonies.

It is also a well-being supplement to people. since there are only tiny quantity available, it can have an increased price than honey which is sold at $3-$5 for a 1 ounce (PS2 or PS3,60 to buy 30 grams).

In order to collect and sell pollen, beekeepers must set up a trap for pollen at the entrance of the beehive.

The harvesting of pollen can be performed for short periods of time, perhaps a couple of weeks per week. This is in order to make sure that enough pollen is available to the honeybees.

PROPOLIS

Propolis, a sticky, resin-like substance that bees use to close cracks and clean the honeybee hive.

Propolis is also a medical supplement which are more effective than pollen. Propolis is available from an beekeeper at $6 or $8 for an one ounce (PS4 up to PS6 equivalent for 30g).

Propolis is removed from the hive with the trap for propolis, and even though small quantities are collected A healthy colony can provide a steady quantity.

Chapter 14: The Royal Jelly

Royal jelly, commonly referred to as bee milk, an incredibly protein-rich, white product produced by bee worker glands.

In the beginning three days of life the larvae of all bees are given royal jelly. The larvae chosen to be queens receive royal jelly all the way through their development. Queen bees consume royal jelly throughout all of their life.

Royal jelly is extremely valuable as it's extremely rare and challenging to find.

It's been utilized over the years to improve health and can be bought for $7 to $8 per one ounce (PS4 or PS6 to get 30 grams).

BEESWAX

A different product of bees that beekeepers are able to make and sell is honey.

It is possible to sell beeswax is or to create value-added products like lip balms or candles.

The cost of beeswax can differ depending on where you reside. However, an average retail cost is around $10 for a kilogram (PS15 each kilogram).

The process of making your own candle can easily be sold for over 20 dollars (PS15) There is an abundance of demand for candles made from beeswax.

VENOM BEE

Bee venom is used for treating arthritis and other ailments is a new concept and is under study.

Bee treatment with stings is one form of apitherapy that involves the use of bees to are able to sting specific parts of the body.

This is an exciting new area of research. If technology advances, it will provide a potential market to beekeepers.

If you are considering offering the treatment of bee stings, then you must consider risks prior to doing so.

Most beekeepers require users of venom to sign a consent form prior to giving any kind of treatment.

2. SERVICES for POLLINATION

The possibility of renting your bees out for pollination can turn into an extremely profitable enterprise.

Many beekeepers earn a livelihood solely from pollination and keep hundreds of beehives on huge farms simultaneously.

They will require lots of equipment, beehives as well as experience.

Smaller farms and horticulture businesses On the other hand often require only the use of a couple of hives for their needs for pollination. This makes it feasible for small bee producers and beekeepers who are part-time to offer pollination services.

The cost you will be charged to hire out bees pollination is contingent upon your area and season.

Farmers typically will pay up 150 dollars (PS108) to pollinate their crops.

3. BEEEKEEPING TOOLS

You can make a variety of products that you can make and offer to other beekeepers, if you are proficient with an electric drill and saw.

Bee feeders, slatter racks entrance reducers escapes and various other equipment which are not included in the original beehive are popular.

Local beekeeper associations can assist to promote your product or if you've got a top-quality product with satisfied clients, it is possible to turn your passion into a profitable business venture.

4. BEKEEPING SERVICES FOR EDUCATION AND CONSULTATION

When you have gained experience as you become more proficient beekeeper, it is possible to share your expertise and earn money through offering training and advice for beginners in beekeeping.

There's plenty to be learned in the initial year of keeping bees, and it may seem difficult to master.

Being able to get someone on hand to offer guidance and training when problems are encountered can be helpful to their success.

Even when you do an active job, you are able to give classes and provide consultation services.

Be sure to schedule beekeeping workshops and demonstrations on weekends. Clients that require your advice will only need a couple of hours per month with you.

After your initial visit with the client after the initial visit, you are able to offer lots of guidance as well as assist via the internet.

5. BEE Elimination

Local services for removal of bees can also be provided by skilled beekeepers. Earning money and growing the size of hives can be achieved at cost to bees.

The price of eradicating bees can range from $150-$1500 (PS108 up to PS1082) or even more if bees are in structures that is difficult to access.

6. Maintenance of the Apiary

Behavioural beekeepers with experience can earn a profit through the keeping of beehives at a cost.

Sometimes, horticulturists or farmers want beehives to assist to pollinate their crops but do not have enough knowledge or the time required to effectively manage them.

They're willing to pay a beekeeper take care of their bees and maintain their hives in top condition.

7. SELLING BEES

Selling bees is a great way to be an effective beekeeper.

You could create substitute bees, start hives as well as replacement bees for different beekeepers by using your beehives.

It can take some time to learn the information and number of bees but when you've mastered it, it could be extremely satisfying.

Here are a few strategies for selling honeybees.

Packages of bees: A package that contains one queen as well as three pounds of bees is priced at about $150 (PS126).

Nucleus colonies, commonly called a nuc is a cluster of honeycombs within a box that includes bees and queen. If you put them in a brand new hive, the current honeycomb gives the bees the chance to get a head-start. Five frames of a nuc that includes a queen can be purchased for between $200 to $250 (PS144 as well as PS180).

Beehives that are established cost from $250 to $350 (PS180 as well as PS252).

How much can you earn from a job in the field of beekeeping?

The sum that you earn from beekeeping is measured by the amount of hives you keep and the quantity of honey you collect.

But, there are many other elements that may affect the earnings you earn, for instance how much hours you are able to work as well as the weather conditions, the location you live in, as well as your ability to keep bees.

You'll need to test and discover the best methods to ensure that beekeeping is profitable for you.

In order to make a income, you'll most likely have to diversify your efforts and utilize a range of different bee products and products and.

Many more opportunities for earning money will be more readily available once you have gained expertise and experience, and you might even be able find a full-time position as a beekeeper at an agricultural beehive.

What are the earnings of beekeepers every year?

Being a beekeeper have the opportunity to earn cash by two methods. First, you can an employment as a beekeeper or become a bee farmer, and collect colonies, keep bees and harvest honey for yourself.

The mean wage for an beekeeper in the United States is $47,899 (PS34,552) annually and an hourly wage of $23 (PS16.60) for an hour as per the Economic Research Institute.

The salary of a beekeeper is a range of $35,523 to $58,231 (PS25,624 to a PS42,000) Based on the your experience and where you live.

The amount of cash that beekeepers make from their hives each year is a lot and contingent on their goals.

Answers to frequently requested questions are listed below, to help you get an idea of how many dollars you can expect making from keeping bees.

Chapter 15: What's The Profit Potential For Each Hive?

If you are honest, be expecting to collect 30-60 pounds (14 to 27 kilograms) of honey from a beehive every year. A healthy and strong colony, on the contrary side, could yield up to 100 pounds (45 kg) of honey per year.

Honey from local sources can range between $5 and 15 dollars (PS3.60 or PS11) according to the region and the quality. It's reasonable to anticipate the price of your honey to be around $10 for a kilogram. (PS15 per kg)

With these figures that you can offer honey at $300- 600 dollars (PS216 up to PS432) per hive over the course of a year.

The sale of other bee products as well as the bees themselves could increase your earnings per beehive.

In order to calculate the profit you earn per hive subtract expenses for maintenance in addition to harvesting packaging and marketing costs.

How Many Bee Hives Earn an Income?

If you're a novice beekeeper and want to earn a all-time source of income, then you're likely to be required to figure out the solution to this concern to you.

Different people have different needs for income or goals in beekeeping and the amount of beehives needed for money varies greatly between one beekeeper and the next.

There are beekeepers who only require just a couple of hives in order to earn their living, as they're skilled marketers who know the importance of retail markets.

Most experienced beekeepers possess a couple of hives, but make money from selling beekeeping equipment as well as providing education.

Others beekeepers are focused on the raising and sale of bees, and creating profitable business in this manner.

If your goal is to earn exclusively from honey production then you'll need a lot of equipment, beehives and space and management of bees.

How many hives will you require when keeping bees to earn money?

How Many Beehives Could One Person Handle?

Once you've got an idea of how much you could earn per beehive, let's take a look at how many beehives an individual will handle.

The solution varies, and is based upon a number of variables, such as age, experience, place and the goals you have.

If you concentrate on the raising of bees to increase honey production, it can be difficult for a single person to handle many honey bee colonies.

Honey production takes longer and needs greater equipment than raising and selling honeybees.

These figures will offer a glimpse of what you can expect in the field of beekeeping to earn money.

For a hobby that is part-time One person could handle up to 25 hives without a lot of planning or control.

Experienced beekeeper, who is focusing on bringing in honey bees for sell, could handle anywhere between 100 and 150 beehives while still working a the full-time work.

If you're managing bees on behalf of others, or leasing your bees to pollinate services, you could also control a large quantity of beehives in part-time.

Beekeepers who are full-time can oversee anything from 400-600 bee colonies, but they will need volunteers for the season to assist in harvesting honey.

Is Commercial Beekeeping Profitable?

Beekeepers that have more than 300 honeybees are termed commercial beekeepers. Their business in beekeeping can be extremely lucrative.

A lot of commercial beekeepers move their colonies in order to offer agricultural producers with pollination and at the same time providing their bees plenty of nectar sources to aid in honey production.

Based on The Grand View Research 2021 Honey Market Analysis The global honey market was worth USD 9.21 billion in 2020. It will likely to expand at a an annual compound growth rate of 8.2 percent.

A high demand for healthy foods has fueled this growth, as consumers become more conscious of the advantages from a healthy lifestyle.

Similar to the reason for this, the natural raw honey can be purchased between $20 and $30 for a pound (PS30 1 kg).

Costs of BEEKEEPING

When you begin beekeeping to earn profits, you should treat it as every other business endeavor.

The first thing you need to consider is the expense of keeping bees.

There are a variety of costs that come with in beekeeping. These include the initial expense of building the hive, cleaning, packing and marketing.

Costs to Start a business

In the context of the advantages of beekeeping as well as earning potential, cost of starting a business in beekeeping is affordable as compared to the other businesses.

A lot of purchases that are also able to endure for years.

Most experienced beekeepers advise that new beekeepers start with two hives during the first year.

This helps you solve some of the problems novice beekeepers encounter and increases your chance of making it through the winter without a colony.

It's also easier for newbies to get started by using nucs, rather than bees, though the cost is higher.

The general rule is that, based on the kind of equipment you choose to purchase as well as the method you use to get your bees, start-up expenses for a beehive fluctuate between $400 to 650 dollars (PS290 or PS470) with a minimum of 300 dollars (PS215) for every additional beehive.

This amount covers the expenses of safety equipment, beehives as well as bees. However, it doesn't include any other additional charges.

Cost of beekeeping for just one hive in the initial year can range from $500-$725 and is inclusive of:

The full hive including every component costs $150-$275 (PS108 as well PS198).

Bee bundles ranging between $125 and $200 (PS90 up to PS145) or nucs from $180-$250 (PS130 up to PS180).

Protection gear can cost from $90-$120 (PS65 or PS87) for each person.

The first year you'll be able to purchase basic equipment at around 45 dollars (PS32) plus other fees at 75 dollars (PS54). Treatments for mites, sugar and pollen for feeding and the books included.

Beekeeping for Newbies offers more information about the cost of starting beekeeping. It also includes an estimate of costs.

Chapter 16: Maintenance Charges

Costs for care of your bees are minimal and fluctuate from year to the following year, based on the weather conditions as well as the supply of nectar and the age of your bees.

Cleaning and fixing hives treatment for mites to protect your honeybees as well as materials for winterizing your beehives, and sugar for feeding the bees when nectar is not available are just some possible expenses for maintenance.

Costs of Marketing and Sales

If you are planning to sell honey, as well as other bee-related products you will need packaging.

It is also important to take into consideration label printing and any flyers that you make for promotions.

Websites and hosting are essential if you want to promote your business or even sell products online.

Renting an stall in a farmer market is incorporated into your marketing and sales fees.

Don't forget to take into account the costs of your time and effort.

REGULATIONS for BEEKEEPING

Prior to beginning beekeeping to make money, be aware of any regulations within your region.

For more information the details, check your department's site and search for "apiary licensure" as well as "beekeeping licence."

There is the possibility that you should verify your local zoning laws or homeowners ' association law.

The answers to these questions are available below.

Do you need to record your the hives?

A few states require that each beehive must be registered as well as an authorization be sought prior to moving bees across state lines. Inspections may also be required.

Are you required to have a commercial license for your region to keep the bees as well as sell their products?

Does product liability insurance have to be purchased at your place of business to set up an apiary business?

Does beekeeping even have a place in the local laws governing zoning?

Are there any good local place to install your hives you do not?

Do you have any local laws restricting the number of hives that you may have on your land? It could be an issue when you live in a suburb or an urban zone.

Certain areas may not permit you to establish a beekeeping company, but they will permit the keeping of bees as an enjoyable hobby.

It is possible to offer honey and other honey-related products for an activity for pleasure.

In order to stay within the limits of your hobby activities it is necessary to restrict the sales you make.

What data should be listed on the label of your honey? There are many states that have specific guidelines for the information of honey labeling. Since you're selling products for consumption by humans There is a lot of regulation from officials of the government.

If you want to have a profitable farm, you should combine the best crops and beekeeping.

the benefits and drawbacks of profitable beekeeping

The benefits of beekeeping are numerous and benefits not only for you but also for the gardens and the surrounding ecological environment. There are many benefits from beekeeping.

1. Natural Honey

One of the greatest benefits of apiculture is the continuous supply of raw honey that is authentic.

2. Beneficial Byproducts

The most popular bee products are available for sale or use comprise beeswax and pollen, propolis and royal jelly.

With such a broad range of products that could be profitable that beekeeping can provide, it is a great option for those who want to earn a profit on their homestead.

3. The Crop Yield has risen

Honeybees play a significant role in pollinating numerous vegetable and fruit crops.

The addition of a beehive into your farm or garden is a great way to increase pollination and the production of your plant.

4. Simple Maintenance

Bees are hard-working animals that need no constant surveillance.

A bee colony that is healthy and is managed properly will produce honey and products made by bees, with little effort from you.

In season of summer, you can devote around an hour per week monitoring and maintaining your colony of bees, but they might require more effort in winter.

5. Many sources of income

There are many ways to make money with beekeeping in several ways, but not only through the sale of honey.

With such a broad range of options, you are able to reach a broad range of prospective customers.

Also, it allows you to look around and discover something lucrative you love doing.

Drawbacks

In light of all the benefits there are, it's hard to imagine the reason why anyone who is a homesteader shouldn't begin the business of beekeeping to earn money. But, just like other livestock, there are a few negatives associated with keeping bees.

1. Initial Capital

The first cost for getting going can be overwhelming when you're short on funds.

2. The first year isn't easy.

The very first year in keeping bees is a long learning curve. You will most likely not be receiving the honey you would like.

Take your time. Bees have to be very active for the first time to establish themselves, and also store honey to last through winter.

3. Infections

Bees are prone to exposure to illness, pesticides, as well as parasites. They need to be monitored and controlled regularly for optimal health and productivity.

It's an excellent idea to meet other beekeepers from your area for information about their issues and ways to help your honeybees while keeping their health.

4. Stings

Bee stings can be uncomfortable and if you're sensitive to bee stings, they could become deadly.

Bees however don't sting until they feel provoked. The chances of stinging are less when you know how to handle your hives effectively and follow appropriate safety measures.

5. Regulations and rules

Regulations regulate the farming of bees as well as the selling of honey at various places.

You may find that you'll have to get your beehives licensed or provide specific details on the honey you harvest. To prevent a possible cost, do your research before you begin your beekeeping.